BASIC
CONTROL SYSTEM
TECHNOLOGY

C. J. Chesmond

Formerly Principal Lecturer in Electronic and Computer Systems
Engineering, Queensland University of Technology

VNR VAN NOSTRAND REINHOLD
New York

Copyright © 1990 by C. J. Chesmond

ISBN 0–442–30386–6

Published in the United States of America by
Van Nostrand Reinhold
115 Fifth Avenue
New York, New York 10003

Distributed in Canada by
Nelson Canada
1120 Birchmount Road
Scarborough, Ontario M1K 5G4, Canada

16 15 14 13 12 11 10 9 8 7 6 5 4 3 2 1

Library of Congress Cataloging-in-Publication Data

Printed in Great Britain

Contents

Preface

Since publishing *Control System Technology* in 1982, there have been many developments in the field of Control Engineering. Many of these have resulted in the progressive elimination of any clear demarcation between this field and those of Computer Engineering, Signal Processing, Artificial Intelligence, Telecommunications, Data Communications, and so on. Thus, in attempting to keep pace with the rapid growth in knowledge which the practising Control Engineer could reasonably be expected to possess, I have been forced to expand the previous material into two volumes.

The present volume, entitled *Basic Control System Technology*, is the first of the two. Its contents are confined to an increasingly detailed description of the basic building blocks of automatic control systems, including transducers, amplifiers, controllers, and final control elements, with new chapters dealing with a basic treatment of such important topics as data transmission, recording instruments and noise reduction techniques.

Material, present in the original monograph, covering such topics as signal and data conditioning, experimental testing, and control system performance and commissioning has had to be carried over into the second volume of the two: this will be entitled *Advanced Control System Technology*. This volume will deal also with the many facets associated with the use of computers as control system elements.

The combination of the two volumes will represent a continued attempt to provide a balanced introduction to the complete field of knowledge which could reasonably be expected to be encompassed by the competent Control Engineer entering the final decade of the twentieth century. What the future holds for such a person in the next century is difficult to speculate upon. The breadth of knowledge which will then be required may be close to the limits of capability of the average engineer, although the creation of appropriate expert systems and the availability of intelligent control systems can be expected to alleviate the situation somewhat.

There is always the danger that engineers trained to an increasing degree of specialisation, such as in the fields of computer engineering or signal process engineering, may be called upon to create complex control systems without having the grasp of the holistic approach to control system design and development which has been achievable until now in the training of Control Engineers. One hopes that these two new volumes, and others like them, will at least demonstrate that a balanced but comprehensive approach to control system design and development is a desirable and hopefully attainable goal.

1

Classification, Terminology and Definitions

1.1 Natural control systems

Feedback control systems exist in nature to a considerable extent. The simple action of a human picking up a pencil typically involves two feedback paths: firstly, visual feedback data enable the current positions of the fingers to be signalled to the optical system and hence to the brain; secondly, having located the pencil, feedback data are transmitted to the brain, via the nervous system, to signal the amount of pressure currently being applied by the fingertips to the pencil.

The objective of the brain is to establish the desired positions for the fingers and the desired degrees of finger pressure, to compare these desired values with the actual values being transmitted back to the brain, and to use the results of these comparisons to compute an appropriate course of action which will then be implemented, by appropriate body muscles, upon receipt of suitable signals from the brain.

Systems such as this, which employ feedback data, are termed *closed loop control systems*. The converse of these are termed *open loop control systems*, such as would result, in the example under consideration, if the person involved were blind and had finger tips insensitive to skin pressure: it is still conceivable that the pencil could be picked up, but the probabilities would be high for missing the pencil altogether, failing to grip it, subsequently dropping it, or snapping it! Thus, with closed loop action there is the potential for considerably improved quality of control compared with open loop action.

The contents of this volume are concerned with automatic control systems, in which the functions alluded to so far are implemented by hardware, with the function of the human operator reduced to the task of establishing desired values, or goals, for the automatic systems.

1.2 Evolution of control systems

One of the first automatic systems to be documented was constructed in pre-Christian times, to open the doors of an ancient Greek temple. The lighting of the fire on an altar caused water to be driven by pressure into a bucket and the resulting additional weight was used to actuate the door-opening mechanism. This was inherently an open loop system, because there were no feedback data supplied to the hardware to indicate the actual position of the doors.

The first significant closed loop control system was James Watt's flyball governor, developed in 1788 for the speed control of a steam engine. Minor developments occurred from time to time (for example in windmills and machine looms), but the real watershed for control engineering was triggered by the Second World War and has continued ever since, accelerated until recently by the space programs. Initial progress was made in single loop systems, which contain a single feedback channel, but the technology has been extended to embrace multi-loop systems, which contain two or more feedback paths: thus, in a modern transport aircraft systems will be present to control automatically such variables as altitude, rate of climb and descent, Mach number, air speed, airport approach trajectory, cruise flight path, etc., in addition to the more straightforward variables such as altitude and rotational velocity, and these are implemented with a hierarchy of feedback control loops.

In the aerospace field, it can be anticipated that there will be progressive improvement in the degree of sophistication of the control systems used. Precision guidance of space probes, the wide range of operational modes of the space shuttle, and the ability, remotely from Earth, to manoeuvre vehicles traversing areas of the surface of planets all require a high level of complexity for the control systems involved.

In the industrial field, repetitive production line operations are increasingly being taken over by robots, which can be designed to operate in the most hostile of environments and which can function for twenty-four hours per day without exhibiting fatigue. Progressive improvement in the control of product quality can be expected to result from improvements in instrumentation hardware.

Control engineering principles can be applied to other fields, such as environments, economies, company management and resource management. Wherever controlled disturbances are applied to a system, of whatever description, in order to improve the performance of that system, then, provided that the performance is monitored and the performance measures are used to determine appropriately the nature of the disturbances, control principles are being used and control theory may be applied to analyse and predict the behaviour of the system.

Because of the rapid evolution of digital computers occurring over the last few years, much of the electronic hardware used in control systems has become computer based and the level of intelligence inherent within the systems has escalated. It has become increasingly difficult to differentiate between the fields of control engineering, computer engineering and data communications, to the extent that soon it may become pointless to attempt such a differentiation.

Note that, although the digital hardware is very different from its analog counterpart, the control principles employed are changed little by this difference.

1.3 Generalised single loop continuous feedback control system

Figure 1.1 is a generalised representation which is valid for any single closed loop continuous control system. Notice that the plant process forms an integral part of the control loop: the implications from this are, firstly, that the performance of the control loop is heavily influenced by the performance of the plant process and, secondly, that the control engineer will need to have a fairly intimate knowledge of the details of the plant process being controlled.

A transducer is a device which is capable of converting the value of a data variable into a signal whose magnitude and sense are representative of the magnitude and sense of the data variable. Typically, the feedback transducer, which is measuring the actual value of the variable being controlled, is mounted direcly on the plant under control, whereas the reference transducer is mounted on a control station (for example, a console) which may be remote from the plant; the reference transducer indicates the desired value (which is typically selected by an operator) of the variable under control.

The error detector performs the function of computing the error signal, which is the difference between the reference and feedback signals and therefore represents a measure of the difference between the desired and actual values for the controlled variable. Obviously, then, a perfect control system is one in which the error signal is held at zero at all times, so that any departure from this condition represents a performance degradation from the ideal.

The final control element is some actuating device physically integral with the plant and which is capable of manipulating the plant in such a way that the controlled variable is, in fact, capable of being adjusted. Because

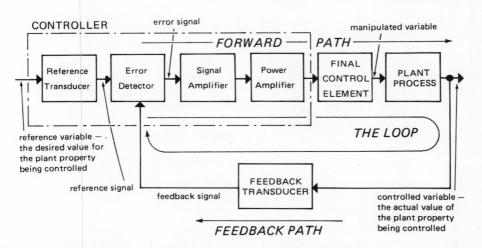

Fig 1.1 Generalised representation of a feedback control loop

of the input signal requirements of the final control element, stages of signal and power amplification are generally necessary in order to boost the error signal strength appropriately.

Typically, the amplifiers and the error detector join the reference transducer in the control station; when mounted together as a single unit they are generally referred to as a *controller*. The controller may be remote from the remainder of the loop elements, even to the extent of being connected by telemetry link in some cases.

Because of the nature of the hardware, the power level at the output from the plant will typically be many orders of magnitude greater than the power level at the input to the reference transducer, so that it may be possible to control MW of output power with mW of input power, for example.

1.4   Classification of control systems

Automatic control systems may be classified in many different ways and these are outlined below. Any one control system will obviously relate to several of the categories listed.

Open loop versus closed loop continuous control

Figure 1.2 represents an elementary example of a simple open loop speed control system, using a separately excited DC motor. The speed of the output shaft will be set manually, by adjustment of the field rheostat, so that, in theory at least, there will be a given specific speed for each position of the rheostat slider, which may, therefore, be calibrated in terms of equivalent shaft rpm.

The accuracy with which a particular desired speed is actually attained will be impaired by the following factors:

● changes in the supply voltages

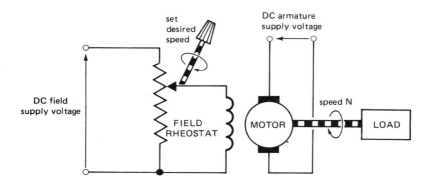

Fig 1.2   Elementary example of a simple open loop speed control system

- variations in the resistances of the rheostat, field winding, and armature winding, resulting from temperature changes due to self heating or fluctuations in ambient temperature
- variations in the characteristics of the load
- magnetic hysteresis in the motor, which will cause the value of speed attained to depend upon the recent past history of variations in desired speed setting.

The degree of precision with which the speed is obtained may be improved considerably, by monitoring the shaft speed with a suitable instrument (a velocity transducer) and requiring a human operator to make appropriate adjustments to the rheostat setting, in order to correct for any drift in the measured speed away from the desired value. This arrangement represents an elementary form of closed loop control, with the operator serving as the error-correcting part of the loop, so that the quality of control will

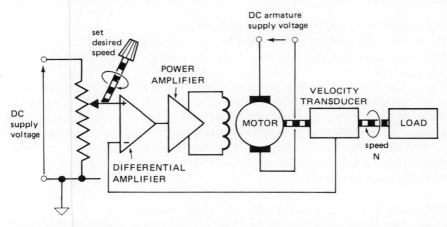

Fig 1.3 Elementary example of a simple closed loop speed control system

depend largely upon the manual dexterity and mental concentration of the operator. The system may be automated, by employing hardware based on the block diagram of Fig. 1.1, so that the quality of control should now be consistent. Figure 1.3 represents a typical closed loop arrangement developed from the open loop configuration of Fig. 1.2.

Generally speaking, the properties of an open loop system are as follows:

Advantages	*Disadvantages*
Relatively simple, resulting in cost, reliability and maintainability advantages	Relatively slow in response to demanded changes
Inherently stable	Inaccurate, due to lack of corrective action for error (that is, departure of actual value from desired value)

Generally speaking, the properties of a closed loop system are as follows:

Advantages	*Disadvantages*
Relatively fast in response to demanded changes	Relatively complex
Relatively accurate in matching actual to desired value	Potentially unstable, under fault conditions

Instability is a condition with feedback systems whereby control is lost and the actual value ceases to track the desired value. Typically, but not inevitably, the output oscillates and these oscillations progressively increase in magnitude. With adequate design, instability can develop only after an equipment failure has occurred. Obviously, the effects can be catastrophic (for example in aircraft or nuclear reactors) so that, in these situations, it is necessary to detect failure and to disable the control system as rapidly as possible: inevitably, this will introduce further complexity.

Classification by type of plant being controlled

This is self explanatory: examples are power generator control, boiler control, air-conditioning control, aircraft control, ship control, space vehicle control, etc.

Where the function of the plant is the manufacture of a product in a more or less continuous production process, the control in this context is referred to as *process control*. In these applications, a wide range of off-the-shelf controllers is available commercially, and these are referred to as *process controllers*. The method of implementation of control, together with the terminology used, has tended to differ from the practice in other areas of application.

Classification by type of process variable being controlled

Again, this is self explanatory: examples are considerable and include displacement, velocity, acceleration, force, torque, tension, temperature, pressure, mass, liquid and gas flow rate, humidity, liquid level, chemical composition, pH, voltage, current, frequency, neutron flux density, altitude, air speed, Mach number, rate of climb and descent, etc.

In a closed loop system, it is desirable that the feedback transducer should provide a direct measure of the variable being controlled. In a limited number of instances, no suitable instrument is available, so that the feedback data have to be generated by computation from measurements by suitable transducers of related process variables. It should be appreciated that a closed loop system is only as good as its transducers, so that accuracy of control cannot be better than the accuracy of the transducers (it will often be significantly worse). It should also be noted that if a process variable cannot be instrumented then it cannot be controlled in a closed loop arrangement; in other words, the feedback transducer should be measuring that variable (or property) of the plant process which one desires to control.

Servomechanisms versus regulators

A *servomechanism* is a closed loop continuous control system in which the plant output is mechanical in its nature; the function of the system is to cause the actual value to track as accurately as possible changes (which may be rapid) in the desired value. Representative examples would be position control systems for aircraft control surfaces (ailerons, elevators, rudders, etc.), robotic arms, nuclear fuel rod loaders, and graph plotters and speed control systems for mine winders and steel mill rollers.

A *regulator* is a closed loop continuous control system the function of which is to hold the actual value at a constant level, determined by a preset desired value (or *set point*) in the presence of fluctuating operational conditions. Representative examples would be automatic voltage and frequency control systems for electric power generators, temperature and liquid level control systems in breweries and oil refineries, and pressure control systems on steam generating plant.

In practice, the distinction between a servomechanism and a regulator can be imprecise, especially where speed control is involved. In any case, the principles involved in the construction, analysis, and design of the two categories of system are identical.

Classification by type of control signals being employed

In this context, the system is defined by the nature of the signals involved and will therefore be electrical, mechanical, hydraulic, pneumatic, or combinations thereof.

Analog, digital, and hybrid control elements

With *analog* elements, the output signal will vary in a smooth, continuous manner when the input signal is varied in a smooth, continuous manner. Most electrical elements and virtually all non-electrical elements are inherently analog.

With a *digital* element, the output signal (or signals) takes the form of a pattern of voltages (or currents) and this pattern (usually in a suitable binary code) is representative of the value of the output data. In the case of *serial* transmission, one data channel is used and the pattern is transmitted as a time sequence of pulses; in the case of *parallel* transmission, there are as many data channels as the code word length (that is, the number of binary digits in the pattern) and all of the bits in the word are transmitted concurrently.

Hybrid systems, which contain some analog and some digital elements, are referred to as *sampled data* control systems. Since almost all plant processes are inherently analog it follows that any continuous control system using digital control elements is a sampled data system.

The majority of electrical analog elements use a varying DC voltage or current as the control signal. In some instances, however, the value of the data being represented is used to amplitude modulate a sinusoidal carrier (typically at 50, 60, 400 Hz, 1, 2, or 4 kHz) and a system using this type of device is known as an *AC-carrier* control system.

Computer control and conventional control systems

A control system may be classified according to whether or not a digital computer is being used in the implementation of the system. In some cases the computer provides the hardware for the controller (in which case this is called *direct digital control*). In other cases the computer does not contribute to but generates data to be used by the loop hardware (whence this is called *supervisory computer control*).

Linear versus nonlinear continuous control systems

A linear control system is one entirely composed of elements which exhibit a straight line relationship between the data value represented by the output signal and the data value represented by the input signal, under steady state conditions of calibration.

A nonlinear control system is one containing one or more elements which do not exhibit such a straight line relationship—examples are shown in Fig. 1.4. It is important to appreciate first that no element is completely

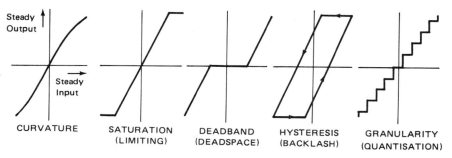

| CURVATURE | SATURATION (LIMITING) | DEADBAND (DEADSPACE) | HYSTERESIS (BACKLASH) | GRANULARITY (QUANTISATION) |

Fig 1.4   Common nonlinear characteristics

linear since, for example, it must be driven to some practical limit if the input signal is sufficiently excessive, and second that most apparently nonlinear elements, in fact, exhibit approximately linear behaviour for relatively small input signal excursions about a nominated quiescent operating level.

Single versus multiple loop continuous control systems

Many systems involve more than one feedback element and therefore contain a multiplicity of feedback loops. Moreover, localised ('minor') feedback loops are often placed around individual system elements or small groups of elements: one function of such a feedback is to 'disguise' any nonlinearity in an element in the forward path of that loop, and this is said to result in a 'linearising' effect.

Sequence control systems versus continuous control systems

A sequence control system is a system of electronic, electrical, or pneumatic digital logic elements, arranged to make on-off decisions at prescribed instants in time. Such a system may be used on its own, with on-off final control elements, in processes involving repetitious operation: feedback may be involved, in the form of on-off signals from limit switches or some other on-off sensor mounted on the plant. Alternatively, the sequence control system may be used (in conjunction with continuous control systems) for performing such functions as controlled start-up and shut-down, up-dating of controller set points, etc.

1.5   Choice of system hardware

It will be relevant to indicate here the types of factor which will need to be taken into consideration when selecting control hardware.

Feedback transducers

For some types of process variable, the range of alternative transducers commercially available may be considerable: examples are displacement, flow, liquid level and temperature instrumentation. In other cases, the range is extremely limited: examples are velocity, humidity, chemical composition, and neutron flux density instrumentation. The possession of a detailed knowledge of the alternatives available is obviously essential for the creation of viable systems. Bearing in mind that the transducer is to be associated intimately with the plant, the factors which could be relevant to the selection of a suitable device are:

- cost
- availability
- ruggedness, in respect to the plant environment
- range
- accuracy
- linearity
- repeatability
- speed of response
- reliability
- maintainability
- life
- power supply requirements
- physical compatibility with the plant
- signal compatibility with the controller
- signal-to-noise ratio.

Chapter 2 covers the principles of transduction and data transmission, whilst Chapters 3, 4, 5 and 6 deal extensively with the more common types of transducer.

In some instances, the output signal may be incompatible with the input requirements of the controller and then it is necessary to interpose signal conversion and/or signal conditioning hardware. Some examples of this would be:

- square-root extractors to compensate for the square law relationship inherent in some flow transducers
- air-to-current converters to enable transducers generating pneumatic signals to be interfaced to electronic controllers
- digital-analog converters to enable digital transducers to be interfaced to analog controllers, and digital controllers to be interfaced to analog final control elements
- analog-digital converters to enable analog transducers to be interfaced to digital controllers
- demodulators to enable AC-carrier transducers to be interfaced to DC controllers
- noise filters to remove parasitic noise from corrupted signals.

Extensive coverage of signal conversion and signal conditioning hardware will be provided in a companion volume entitled *Advanced Control System Technology*.

Final control elements

To a major extent, these are either control valves (globe or butterfly types, with pneumatic, electric, or hydraulic actuators), heaters, or motors. The factors which would be relevant to the selection of a suitable motor, for example, are:

- cost
- availability
- ruggedness, in respect to the plant environment
- load details: inertia, friction constants, torque loadings
- maximum and minimum velocity
- maximum acceleration
- duty cycle
- reliability
- maintainability
- life
- mounting and coupling requirements
- power supply requirements
- input signal characteristics.

Chapters 8, 9 and 10 describe the commonly used types of final control element.

Controllers

These complete the feedback loop, accepting the feedback signal as an input and generating, as an output, the input signal required by the final control element. The controller may be custom built or it may be an off-the-shelf commercial item.

Particularly in the custom-built cases, the production of the controller may involve considerable design, development, manufacturing, and commissioning effort. The performance of the plant process, feedback transducer and final control element is, to a large extent, predetermined for a particular situation, so that it is necessary to have flexibility in the controller characteristics, in order to obtain satisfactory performance from the completed control loop.

Chapters 8 to 12 inclusive cover the wide range of hardware likely to be encountered in both custom-built and off-the-shelf controllers. In addition Chapter 3 contains descriptions of the most common reference transducers.

The companion volume will provide much additional information on controller details, particularly where the application of digital computers as controllers is concerned.

Chapter 13 presents a range of examples of complete control loops, with schematic diagrams showing how transducers, controller hardware, final control elements and the plant process will interact within each loop.

Chapter 7 provides details of the instruments most commonly used for recording the performance of control systems.

2

Principles of Transduction and Data Transmission

2.1 Introduction

Before considering, in Chapters 3 to 6, the principles and physical details of the most commonly used feedback and reference transducers, it is necessary to define the terms to which reference is often made in relation to these devices. Following that, we shall consider in detail all of the principles which might need to be followed when installing transducers into systems, bearing in mind that feedback transducers will always be exposed to the plant environment, which will often be hostile, and that reference transducers may also be similarly exposed, in some instances. The chapter will end with a description of the techniques commonly used when installing transducers: these amount to the exercising of 'good practice' which is particularly important when a transducer is physically placed at a significant distance from the device being driven by its output signal(s).

2.2 Definitions of terms used in transducer specifications

Transducer A transducer is a device or combination of devices whose response to a mechanical, electrical, physical or chemical stimulus is an output signal functionally related, in terms of magnitude and sense, to the magnitude and sense of the stimulus. The transducer may often be composed of two principal components: the *primary element*, or *sensor*, which is that component in direct contact with the source of the stimulus, and the *secondary element*, which is that component directly responsible for the generation of the output signal. Sensor and transducer stimuli generally fall into one of the following six groups.

Physico-mechanical Force, weight, displacement (rectilinear or angular), velocity (rectilinear or angular), acceleration (rectilinear or angular), pressure, differential pressure, flowrate (linear or volumetric), level, hardness, thickness, viscosity, density, colour, opacity, turbidity, moisture content, particle concentration, vibration, time, etc.

Electrical   Voltage, current, frequency, phase angle, power, reactive volt-amperes, inductance, capacitance, resistance, energy, etc.

Magnetic   Field strength, hysteresis, coercivity, induction, etc.

Thermal   Temperature, thermal conductivity, heat flowrate, emissivity, etc.

Radiation   Radio frequency, infrared, visible, ultraviolet, X ray, nuclear, etc.

Chemical   Concentration, composition, etc.

Transmitter   A transmitter is a transducer or secondary element whose output signal takes a standardised form directly suitable for transmission to another device.

Converter   A (transducer) converter is a transducer which converts a signal of one standardised form into a signal of a different standardised form.

Accuracy   The accuracy of a transducer is the closeness to which the output signal approaches the true value which it should have for any specific value of applied stimulus. Many factors contribute to inaccuracy (error) in a transducer—some of these will be discussed in the following chapters.

Sensitivity   Sensitivity is the ratio of the change in output signal to the change in the value of input stimulus causing the change in output. Thus, if a 1 degree change in input angle of an angular displacement transducer results in a 2 V change in its output signal, it is said to have a 2 V/degree sensitivity.

Range (span)   The range or span of a transducer is the difference between the maximum value and the minimum value of the input stimulus to which the output signal responds. If the minimum value is non-zero then the ratio (maximum input stimulus value)/(minimum input stimulus value) is referred to as *rangeability*.

Linearity error   If the sensitivity of a transducer over its entire range is precisely constant then it may be considered to be *linear*. However, no practical device can be perfectly linear and departure from perfect linearity can be defined in a number of alternative ways, as shown (in exaggerated form) in Fig. 2.1.

Note that a supplier may well use whichever definition of linearity error yields that most favourable figure, but will not necessarily state which definition has been used. Note also that, with some transducers, the loaded linearity error may be significantly worse than the published

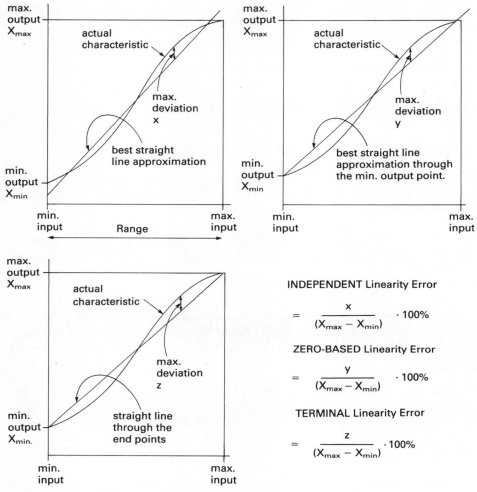

The INDEPENDENT Linearity Error

$$= \frac{x}{(X_{max} - X_{min})} \cdot 100\%$$

ZERO-BASED Linearity Error

$$= \frac{y}{(X_{max} - X_{min})} \cdot 100\%$$

TERMINAL Linearity Error

$$= \frac{z}{(X_{max} - X_{min})} \cdot 100\%$$

Fig 2.1   Alternative definitions of linearity error

(unloaded) value: if loading does cause unacceptable errors then some form of buffer amplifier may need to be interposed between the transducer and the load.

Resolution   Some transducers are incapable of sensing very small changes in input stimulus. Resolution is defined as the smallest change in input stimulus to which the output will respond, expressed as a percentage of the input range. Thus, a rectilinear displacement transducer with a range of 100 cm and a resolution of 0.1% will be incapable of sensing changes in input displacement of less than 1 mm. The presence of resolution error results in a small 'staircase' on the output vs input characteristic; this phenomenon is known as *quantisation* or *granularity*. Where a transducer can sense infinitesimally small changes in stimulus it is said to have *infinite resolution*.

Precision   Precision is a measure of the degree of *repeatability* of successive measurements of the same value of input stimulus. Phenomena such as hysteresis and stiction can account for precision errors. The distinction between precision and accuracy should be noted: figures for the two (and for resolution) may be significantly different for a particular device. A digital voltmeter may equally consistently read a true voltage of 5.000 V as either 4.996, 4.997 or 4.998 V: the accuracy is 3 parts in 5000, the resolution is 1 part in 5000, whilst the precision is 2 parts in 5000.

However, in many transducers precision and resolution may be synonymous, if they are both accounted for by the same causes: in transducers involving mechanical parts, stiction and mechanical hysteresis will limit repeatability and will also place limits on the degree of resolution.

2.3 Common sources of electrical noise

Any unwanted, parasitic component of signal occurring within a circuit can be regarded as 'noise', if the circuit is potentially capable of responding to that signal. The magnitude of the noise relative to that of the 'useful' component of signal is expressed in terms of *signal-to-noise ratio* (SNR). SNR = 20 log [(RMS voltage of useful signal)/(RMS voltage of noise)] and is a dB figure.

Noise may take many forms:

- a steady DC component
- a steady alternating (AC) component, which is not necessarily sinusoidal
- a steady pulse train
- a randomly varying component
- a transient component, which may be a varying DC, AC, or pulse train only present temporarily.

Typical sources of parasitic DC voltages include:

- rectified parasitic AC voltages
- saturation of amplifiers resulting in waveform distortion
- galvanic action due to the presence of moisture or water vapour at the interface between two dissimilar metal surfaces resulting in the establishment of a chemical wet cell
- electrolytic action, due to the presence of (even slightly) acidic or alkaline moisture between the surfaces of two metals which need not be dissimilar.

Typical sources of parasitic AC voltages include:

- AC mains supplies (mains frequency)
- radio and TV transmissions (high frequencies)
- instability due to positive feedback (high frequencies).

Typical sources of pulse trains include:

- trigger circuits for SCR and triac networks (see Section 8.4)
- digital communication channels.

Typical sources of randomly varying voltages include:

- intermittently poor connections
- vibration of conductors in magnetic fields or of conductors possessing voids in their insulation material
- thermal agitation of electrons within a resistance (thermal noise) defined by the law
open-circuit noise voltage $= (4kTBR)^{1/2}$ where
    $k = $ Boltzmann's constant 1.38×10^{-23} J/K
    $T = $ absolute temperature in K
    $B = $ noise bandwidth in Hz
    $R = $ resistance in ohms
- arcing at electrical contacts, such as the brushes on DC machines.

Typical sources of transient components include:

- switching of power sources
- switching of loads (especially inductive loads) on power supplies
- lightning strikes.

2.4   Elimination of noise at the source

As an alternative to the reduction of the effect of noise in the circuit being affected, it may sometimes be practicable to reduce the noise at its source. The techniques to be used would be many and varied reflecting the considerable variety of types of noise source. The following list indicates some of the alternative techniques which might sometimes be used:

- rescaling of signals or the replacement of amplifiers with others having higher output swings, to eliminate saturation
- sealing of electrical joints against the ingress of moisture
- the use of filters to attenuate AC voltages and transients
- strapping down conductors subjected to vibration

- the use of arc suppression and surge suppression networks
- physically resiting the noise source
- placing electromagnetic (ferrous) or electrostatic (earthed) screens around noise sources.

2.5 Mechanisms causing noise introduction

There are eight alternative basic mechanisms which can result in the introduction of noise into a circuit.

Inductive coupling   A current flowing in a conductor will establish an electromagnetic field around the conductor. If a conductor in the circuit subjected to induced noise is situated within that field then an EMF will be induced whenever the original current changes. The two conductors, together with the medium separating them, form a mutual inductance. The magnitude of the current, which consequently may be circulated in the circuit by the induced EMF, will depend partly upon the impedance of the circuit presented to the EMF. This phenomenon, often called 'inductive coupling', tends to favour low-frequency noise.

Capacitive coupling   Whenever two conductors are in close proximity and a potential difference exists between them, the conductors become electrodes in a capacitor with the intervening space forming the dielectric. Whenever the voltage on one conductor changes the charge on the second conductor will change, so that a current may flow in the circuit of which the second conductor is a part. The value of voltage developed by the current flow will depend upon the impedance of the circuit. An electrostatic field exists between the two conductors and the resulting phenomenon is often called 'capacitive coupling'; it tends to favour high-frequency noise.

Common impedance coupling   Whenever two sources of voltage circulate currents through a common impedance a current flowing in one loop will result in a voltage being created in the other loop, as shown in Fig. 2.2.

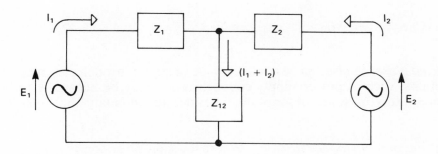

Fig 2.2 Circuit with two loops sharing a common impedance

In this circuit

$$I_1 = \left(\frac{1}{Z_1 + Z_2}\right) E_1 - \left(\frac{Z_{12}}{Z_1 + Z_{12}}\right) I_2$$

$$I_2 = \left(\frac{1}{Z_1 + Z_2}\right) E_2 - \left(\frac{Z_{12}}{Z_2 + Z_{12}}\right) I_1$$

Thus, the value of I_2 is influenced both by E_2 and I_1, so that voltages in the I_2 loop will be influenced by I_1. If Z_{12} represents the impedance of a conductor shared by two circuits then there will be 'crosstalk' induced between the two circuits as a result. It follows that common conductors should be avoided wherever possible.

Earth loops   Where signal common conductors are earthed at several points this results in one or more closed loops being formed, as shown in Fig. 2.3. Any parasitic voltage induced in the signal common conductors will be presented with a low impedance circuit, so that it may circulate a relatively large current around the earth loop.

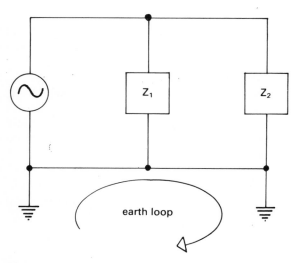

Fig 2.3   Circuit in which an earth loop exists

Conducted noise   Noise can be conducted into a circuit through electrical connections, such as power supply connections. It may be necessary to introduce filters of some description, in order to attenuate the noise signals.

Inadequate common mode rejection   Where the same noise component of voltage exists at the two input terminals of a differential-input amplifier (see Section 8.2), ideally no noise component will be propagated in the amplifier output voltage. In an amplifier with a low figure of 'common mode rejection ratio', a noise component will be generated at the output.

Noise components   Obviously, whenever a component exhibiting an internally generated noise voltage is used in a circuit that voltage will be impressed upon the circuit.

Instability   In feedback loops around high gain active elements there is always the possibility of instability arising. This normally takes the form of high frequency oscillations, the waveforms of which may or may not be sinusoidal. Special compensation networks may be necessary in order to eliminate the instability.

Many alternative techniques are available for preserving a high SNR value. Some of the more common methods will be covered in Section 2.6. These are particularly important in circuits where the level of the useful signal is low: thermocouple and strain gauge analog networks, for example, and most digital networks.

2.6   Techniques for minimising the noise introduced into a circuit

For relatively long interconnections between devices the following techniques can be employed.

- Use, twisted together, a pair of conductors for outgoing and return connections to a load, to minimise electromagnetic 'pick-up'. These twisted pairs are much more effective than straight twin parallel conductors.
- Screen conductors or conductor pairs.
- Do not use the screen as a signal return path: the screen should be earthed at one point only, at the signal source end of the connection.
- Avoid sharing conductors when several loads have to be connected to the same signal source.
- Use low-impedance signal sources.
- Where practicable transmission should be balanced to ground, as shown in Fig. 2.4.
  Common mode EMFs induced into the two conductors should be cancelled out by the differential input action of the load.
- Physically isolate signal conductors from power conductors; where they must cross arrange for them to do so at right-angles.
- Where signal commons of several networks or devices have to be coupled together, avoid a 'daisy-chain' type of connection; the signal commons should all be connected to the one node, which may then be

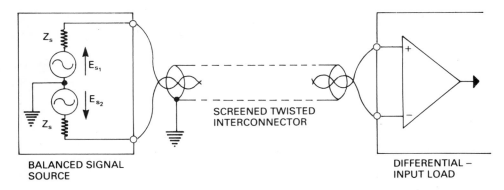

Fig 2.4   Use of balanced transmission to minimise induced noise

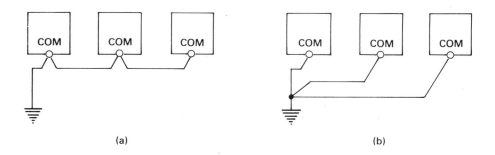

Fig 2.5   Alternative techniques for joining multiple signal commons, showing (a) bad practice and (b) good practice

earthed as necessary (see Fig. 2.5). Multiple cable screens may also be connected to ground using the same type of common node configuration.

Within networks and devices the following techniques can be employed.

- Wherever possible avoid using components which are likely to generate noise.
- Use surge suppression networks on relay contacts.
- Use earthed electrostatic screens on transformers.
- Use ground planes on printed circuits: a ground plane is a large printed copper area on the printed circuit which should be earthed but not used as a conductor. With multilayer printed circuits, the ground plane may be a copper layer overlaying the complete circuit surface area.
- Filter power supplies as heavily as possible. This can involve using very high capacitor values: for example 10 000 μF electrolytics.

- Place bypass capacitors (typically 0.1 μF ceramics) at frequent intervals as close as possible to integrated circuits: the capacitor is connected between the supply rail and signal common. The capacitor provides a short-circuit path to signal common for high-frequency noise voltages. These voltages may be due to induced or conducted noise, or due to instability arising from positive feedback paths being created by the supply rail interconnections.
- Ensure that operational amplifier networks are adequately compensated against high-frequency instability.
- Where a *guard* terminal is supplied on a device, connect it to earth in the manner shown in Fig. 2.6. This terminal will have been connected internally to ground planes and conductor screens.

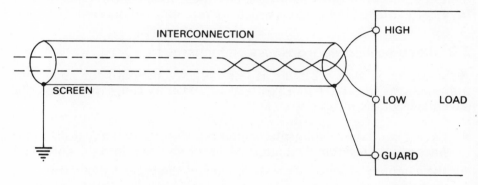

Fig 2.6   Correct method of connection of the guard terminal on a device

2.7   General data transmission requirements

In most plants, the feedback sensors are frequently situated in locations which are remote from the control rooms in which the controllers reside. In most instances, the output signals are quite unsuitable for direct transmission of data to the control room for the following reasons.

- the signal is of an inappropriate nature (for example a differential liquid pressure associated with an orifice plate); and/or
- the signal is of an unsuitably low range such as to make the SNR unacceptably poor with direct transmission (for example the mV range of output signal from a thermocouple); and/or
- the signal is of an unsuitably low range such as to be heavily attenuated by signal losses in the transmission circuits; and/or
- the signal is associated with an unacceptably low power level such as to result in the load 'dragging down' the output signal.

For a combination of these reasons, it is commonplace to use transmitters in order to convert the sensor output signals into signals appropriate for data transmission. Sections 5.1 and 5.2 provide further explanation about transmitters.

The output signals from these transmitters usually take forms from the following alternatives:

- pneumatic signals
- analog electrical signals, which may be DC currents or voltages
- digital electrical signals, usually serial in nature.

The ranges of analog signals conform to industry standards, and the same signal ranges are used for the outputs from controllers, which are used to drive the final control elements. Again, these final control elements are frequently situated in locations which are remote from the control rooms.

2.8   Pneumatic data transmission techniques

As stated in Section 5.2 there is one industry standard for pneumatic transmissions signals: 3 to 15 psi (20 to 100 kPa) gauge pressure. The advantages of the offset datum are:

- having lines above atmospheric pressure ensures that any leaks in the lines cannot result in the ingress of dirt or moisture into the lines; and
- the complete absence of pressure signifies the breakage of a transmission line.

Where the pneumatic transmitter has to supply two or more loads, these loads will be connected in parallel. In those cases where the number of loads is too great for the volumetric capacity of the transmitter, and/or the transmission path is too long for the pressure drop to be acceptable, one or more air relays may be inserted at the intervals in the line in order to sustain pressure levels. See Section 9.1.2 for details on air relays.

2.9   Analog data transmission techniques

As stated in Section 5.2, the most common industry standard for analog electronic transmission is the 4 to 20 mA DC signal. This requires the output of the transmitter to be configured as a current source, as shown in Fig. 2.7. Because of the nature of current sources the following considerations apply:

- the power supply may be sited at the load end of the transmission circuit for convenience
- multiple loads should be connected in series and all but one must be floating
- unused transmitter outputs should be short-circuited

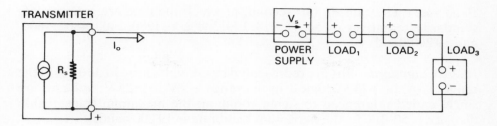

Fig 2.7   Circuit arrangement for DC current transmission

- the total load resistance presented to the transmitter can be in a range from 0 ohms to a specified maximum limit, which will be somewhat less than $V_s/20$ kΩ.
- where a load is voltage sensing, its input terminals may be shunted by a suitable resistance: for example 250 Ω for a 1–5 VDC input range.

Where transmitters use solid state technology, rather than electromechanical techniques, it may become increasingly common for the output to be configured as a voltage source. Because of the nature of voltage sources the following considerations will then apply:

- the power supply must be connected separately to the transmitter
- multiple loads must be connected in parallel and none may be floating
- unused transmitter outputs should be open-circuited
- the total load resistance presented to the transmitter must lie in a range between a specified minimum limit and infinity
- where a load is current sensing, a voltage-current converter will need to be interposed between its input and the transmission signal.

2.10   Digital data transmission techniques

Because of the large number of conductors which would be involved, parallel data transmission is not often used for anything but short distances.

Serial data transmission between digital devices can be specified by means of a number of alternative terms.

Simplex   This involves 2-wire transmission in one direction only.

Half-duplex   This involves 2-wire transmission in both directions, one direction at a time.

Full-duplex   Normally this involves four wires for transmission in both directions simultaneously, two wires for one direction and two for the other.

Baud rate   The rate, in bits/second, at which the data are transmitted; standard alternatives are 50, 75, 110, 150, 300, 600, 1200, 2400, 4800, 9600, and 19 200 baud.

RS232C standard   Bits are represented by DC voltages—logic 0 lies in the range +5 V to +25 V, logic 1 in the range −5 V to −25 V; voltages are single ended referenced to signal common; the maximum transmission distance is 50–100 ft, the maximum baud rate is 19 200 baud.

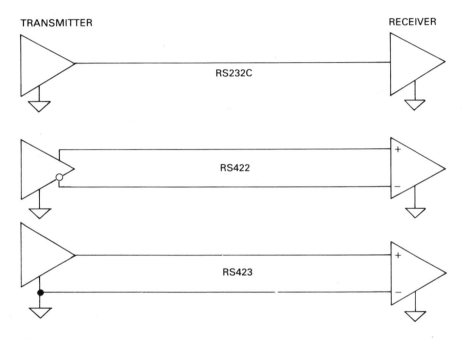

Fig 2.8   Configurations for alternative transmission standards

RS422/423 standards   Bits are represented by DC voltages—logic 0 lies in the range +0.2 V to +6 V, logic 1 in the range −0.2 V to −6 V; with RS422, voltages are transmitted double ended (balanced about signal common) and the receiver is differential input; with RS423, voltages are transmitted single ended (referenced to signal common) and the receiver is differential input (see Fig. 2.8); the maximum transmission distance is 4000 to 5000 ft; baud rates approaching 10^7 baud are possible.

Current loop   Bits are represented by DC currents, the transmitter being configured as a current source and the receiver as a current sink; logic 0 is represented by 0 mA, logic 1 by 20 mA; the maximum transmission distance is 2000 ft.

Asynchronous transmission   A data word is transmitted as a burst of pulses occurring at a predetermined bit rate but commencing at no particular moment in time; each word is preceded by a start bit and concluded by one or two stop bits, for identification purposes.

Table 2.1   ASCII Code

Bit positions 7, 6, 5								Bit positions			
000	001	010	011	100	101	110	111	4	3	2	1
NUL	DLE	SPACE	0	@	P		p	0	0	0	0
SOH	DC1	!	1	A	Q	a	q	0	0	0	1
STX	DC2	"	2	B	R	b	r	0	0	1	0
ETX	DC3	#	3	C	S	c	s	0	0	1	1
EOT	DC4	$	4	D	T	d	t	0	1	0	0
ENQ	NAK	%	5	E	U	e	u	0	1	0	1
ACK	SYN	&	6	F	V	f	v	0	1	1	0
BEL	ETB	'	7	G	W	g	w	0	1	1	1
BS	CAN	(	8	H	X	h	x	1	0	0	0
HT	EM	)	9	I	Y	i	y	1	0	0	1
LF	SUB	*	:	J	Z	j	z	1	0	1	0
VT	ESC	+	;	K	[	k	{	1	0	1	1
FF	FS	,	<	L	/	l	|	1	1	0	0
CR	GS	–	=	M	]	m	}	1	1	0	1
SO	RS	.	>	N	∧	n	~	1	1	1	0
SI	US	/	?	O	—	o	DEL	1	1	1	1

NUL = All zeroes
SOH = Start of header
STX = Start of text
ETX = End of text
EOT = End of transmission
ENQ = Inquiry
ACK = Acknowledgement
BEL = Bell
BS   = Back space
HT   = Horizontal tab
LF   = Line feed
VT   = Vertical tab
FF   = Form feed
CR   = Carriage return
SO   = Shift out
SI   = Shift in
DLE = Data link escape

DC1 = Device control No. 1
DC2 = Device control No. 2
DC3 = Device control No. 3
DC4 = Device control No. 4
NAK = Negative acknowledgement
SYN = Synchronous idle
ETB = End of transmitted block
CAN = Cancel
EM   = End of medium
SUB = Start special sequence
ESC = Escape or break
FS   = File separator
GS   = Group separator
RS   = Record separator
US   = Unit separator
DEL = Delete

Note   Bit 7—most significant bit   Bit 1—least significant bit
Most significant bit is transmitted first

Synchronous transmission   Data words are transmitted as a stream of pulses occurring at a predetermined bit rate and commencing at a specific moment in time; no start or stop bits are transmitted but, periodically, synchronising characters are transmitted in order that the receiver may use them to maintain synchronism between its clock and that of the transmitter.

Modems   When a voice-grade telephone line forms part of a communications channel, the signals are modulated and demodulated by 'modems' situated at the transmitting and receiving ends, respectively, of the line; frequency shift keying (FSK) frequency modulation is used for baud rates not exceeding 300; phase modulation is used for baud rates in excess of 300.

ASCII code   This is the most commonly used code for asynchronous transmission; it uses seven bits to encode 128 possible characters and functions; Table 2.1 indicates the allocation of codes and the abbreviations used.

3

Transducers –
Displacement, Reference and Velocity

3.1 Introduction

The range of devices available to provide transduction of displacement (rectilinear or angular) is considerably greater than that for any other variable. Moreover, many of these displacement measuring devices form the secondary element in transducers for other variables, such as temperature, force, torque, acceleration, tension, liquid flowrate, liquid level, liquid density, etc.: in this role, the function of the displacement measuring device is to enable the transducer to generate an electrical output signal.

Representative applications of displacement and velocity transducers as primary feedback transducers in closed loop control systems would include the following:

- recording instruments
- machine tool position and speed control
- steering of tracking aerials (antennae) for telecommunications, weaponry, and astronomy
- paper mill and steel mill control
- industrial robot control
- control and guidance of aircraft, satellite launchers, missiles, ships and submarines
- control of transportation vehicles.

Because of the considerable range of devices in these categories, space will permit discussion only of the more commonly used types. In the measurement of motion, it is necessary to make a distinction between rotational and translational movement: transducers for measuring rotational motion are said to be 'rotary' or 'angular', whilst those for measuring translational motion are said to be 'rectilinear'. The last named word is preferable to 'linear' when describing the type of motion, because the word 'linear' is best reserved for describing the calibration graph for the transducer. Thus, linear is the antithesis of nonlinear, whilst rectilinear is the antithesis of rotary: a transducer may be rectilinear and nonlinear; alternatively, it may be rotary and linear!

Displacement transducers, which are also known as *position transducers*, are used primarily as feedback transducers in position control systems, which are called, alternatively, *servomechanisms* or *servosystems*. However, other versions of these may be used as reference transducers for a wide variety of control systems, especially where the reference variable value is to be set manually.

Velocity transducers, which are also known as *rate transducers*, are used mostly as principal feedback transducers in velocity (speed) control systems and as secondary feedback transducers in position control systems. In the latter case, the velocity transducer signal is a measure of the rate of change in the displacement transducer signal, and is used principally to modify the dynamic behaviour of the position control system, a process known as 'damping'.

3.2   Displacement transducers

3.2.1   Servo potentiometers

Servo potentiometers are distinguished from conventional potentiometers in that they are manufactured to far superior specifications; these can involve:

● a close tolerance on the linearity of the wiper voltage versus wiper displacement characteristic

● a close tolerance on the resistance of the track

● the use of high-grade bearings, to minimise friction

● the use of a high-quality precious metal brush for the wiper, to minimise friction, wear, and contact resistance.

The track of a servo potentiometer may be wirewound or it may be manufactured from homogeneous material such as carbon composition, high conductivity metal or metal composition film, or conductive plastic. Many servo potentiometers are wirewound, because of the high degree of linearity which can be obtained with this type of construction; however, a disadvantage which results is a granularity effect, due to the fact that the voltage at the wiper must increment by an amount equal to the voltage dropped across one turn of the winding, when the wiper is moved. The wirewound construction also lends itself to the addition of fixed tappings, the use of which is discussed in Section 12.2.7.

The law relating wiper voltage to wiper displacement is sometimes made deliberately nonlinear: one method for achieving this is to graduate the cross-sectional area of the track and, with wirewound types, a stepped form of nonlinear function can also be achieved by using wire of different diameter along different sections of the track.

Rectilinear potentiometers are constructed with lengths of travel ranging typically from 1 mm to 6 m (with 1 cm to 15 cm common), track resistances ranging from 20 Ω to 200 kΩ (with 100 Ω to 10 kΩ common), and linearity errors ranging from less than 0.1% (with 0.1%, 0.5%, and 1% common). Typical resolution for wirewound potentiometers ranges from 70 to 200 steps per centimetre. More than one track may be mounted within the case of the device, in order to provide multiple data channels. Rectilinear potentiometers will be destroyed if overdriven mechanically, so that mechanical limits must be placed externally on the input displacement. Life of these potentiometers may be short, due to the techniques used to maintain a good contact between the wiper and the track.

Rotary potentiometers are manufactured in single-turn and multiturn versions. Single-turn types use a toroidal former for the track and the usual need for a deadspace, to provide electrical isolation between the two ends of the track, means that the useful input travel is less than 360° and typically ranges from 320° to 355°. (Note, however, that Section 12.2.7 describes a toroidal potentiometer which represents an exception to this situation.) Usually, no mechanical limits to travel are included in the device, so that it will not be damaged if the wiper should be driven into the deadspace: if such limits are not incorporated into the input drive then the feedback signal generated by the potentiometer will be lost, whenever the wiper is driven into the deadspace. When this occurs, the load connected to the wiper terminal will be presented with an open circuit signal source.

The typical track resistance range is from 50 Ω to 200 kΩ, with linearity errors extending down to 0.01% and resolution (for wirewound types, and dependent upon the diameter of the toroidal former) extending down to 0.04%. Toroidal potentiometers may be 'ganged' together on a common shaft, to provide multiple data channels.

Multiturn potentiometers use a helical former for the track, and the wiper assembly is designed so that the brush simultaneously rotates and advances axially, as the input shaft is rotated. These potentiometers are made with as many as 25 shaft revolutions for full travel, with 5 and 10 revolutions being the most common. They will be destroyed if overdriven mechanically, so that external mechanical limits must be incorporated into the input drive. The typical track resistance range is from 10 Ω to nearly 1 MΩ, with linearity errors extending down to 0.02% and resolution (for wirewound types, and dependent upon the number of shaft revolutions for full travel) extending down to less than 0.01%. Multiturn potentiometers are sometimes ganged together on a common shaft, to provide multiple data channels.

Figure 3.1 shows the most common method of connection for a servo potentiometer, which is connected as a potential divider supplied from a stable voltage reference source, which is usually DC but occasionally may be AC. In some applications, the voltage supply will be distributed symmetrically about signal common and, in this case, it is preferable to use a centre tap on the potentiometer track and to connect this tap to signal common, thus establishing a rigid electrical datum at the mid-point of the shaft travel: this mid-point would then also be used as the mechanical datum, for calibration purposes.

Figure 3.1 also shows the distorting effect that the presence of significant load current can have on the calibration of the potentiometer, with the deviation between the loaded and unloaded characteristics being greatest at mid-travel: loading can readily degrade the linearity of a potentiometer, so that, where necessary, a precision buffer amplifier should be inserted between the wiper and the load, in order that negligible current will be drawn from the wiper.

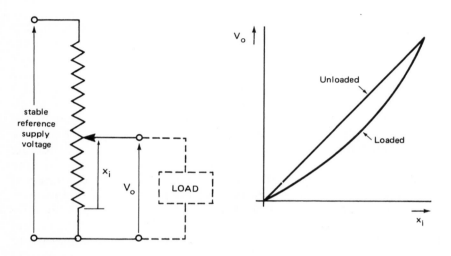

Fig 3.1   Circuit and electrical characteristics of a servo potentiometer

The merits and demerits of potentiometers, in comparison with alternative displacement transducers, are seen to be as follows:

Advantages	*Disadvantages*
Relatively inexpensive	Limited life
Rectilinear and rotary versions are available	High breakaway force/torque
The law can be modified, if tappings are available	Uneven wear of the track will result in a degrading of linearity
DC or AC operation	Poor brush contact will result in parasitic noise generation
Requires no special power supply, usually, provided that the available supply is stable	Relatively low reliability, due to the use of electrical contacts moving over a relatively rough surface
High sensitivity	Limited travel, with the possibility of destruction, if overdriven
Reasonable accuracy	Sensitive to loading effects
Available in a wide range of resistance values	

3.2.2 Differential inductors and transformers

Increasingly, these devices are replacing potentiometers, mainly because of the superior reliability associated with the former. They can be subdivided into four categories:

LVDI – Linear Variable Differential Inductor
LVDT – Linear Variable Differential Transformer
RVDI – Rotary Variable Differential Inductor
RVDT– Rotary Variable Differential Transformer

The LVDI and LVDT are rectilinear transducers, whereas the RVDI and RVDT are rotary. The rotary versions have a very limited travel (typically $\pm 60°$) and are sometimes referred to as 'Rotary Pick-Offs'. The rectilinear versions are by far the more extensively used, mainly because they are available with a wide range of travel. The inherent reliability associated with these inductors and transformers arises because they do not involve the use of any moving electrical contacts: in fact, there need be no physical contact whatsoever between their moving and stationary component parts.

Figure 3.2 shows the electrical circuits for half bridge and full bridge differential inductors and for differential transformers. In each case, the applied AC voltage is sinusoidal and has a frequency which may lie between 50 Hz and 20 kHz but which is usually in the 1 kHz to 5 kHz range. In the case of the differential inductor, the device is connected in an AC Wheatstone bridge arrangement, which is balanced electrically at the null position for the core(s), so that $V_0 = 0$; as the cores are offset, by input displacement x, the self and mutual inductances of the windings vary, so that the bridge becomes unbalanced and V_0 changes. In the case of the differential transformer, the two identical windings are connected in series-opposition so that, when the core is in the null position, $V_1 = V_2$ and $V_0 = 0$; when the core is offset by input displacement x, the magnetic coupling with the primary increases for one secondary and decreases for the other, so that V_0 changes.

Figure 3.3 indicates the type of construction used for the LVDT and RVDT; comparable configurations are used for the LVDI and RVDI respectively. As can be seen, the moving member needs make no physical contact with the fixed member, since magnetic flux linkages form the transducing medium.

Figure 3.4 shows the type of output voltage versus input displacement characteristic obtained with all of these devices. Typically, V_0 will nominally be in phase with V_{AC} for positive values of x, and nominally in antiphase with V_{AC} for negative values of x. In practice, due to the effects of magnetising current, core losses, and winding impedances, there will be small parasitic phase shifts present, so that the phase of V_0 will not be precisely $0°$ or $180°$ relative to V_{AC}. In addition, the presence of ferrous material in the magnetic circuit will cause a small amount of harmonic distortion in the output voltage waveform.

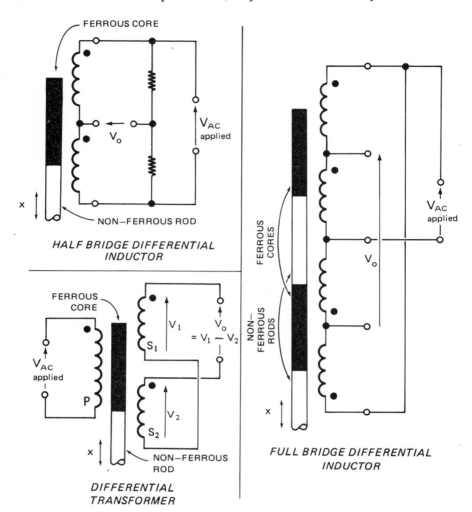

Fig 3.2 Circuits of differential inductors and transformers

The rectilinear types are constructed with travels ranging typically from 1 mm to 30 cm, although this range has been extended to 0.1 mm to 250 cm in certain devices. Linearity errors as low as 0.1% are achieved relatively easily, whilst the resolution is infinite, because the V_0 vs x characteristic is stepless. The advantages with these transducers are seen to be as follows:

- rugged
- zero breakaway force/torque
- stepless characterstic
- relatively insensitive to loading effects

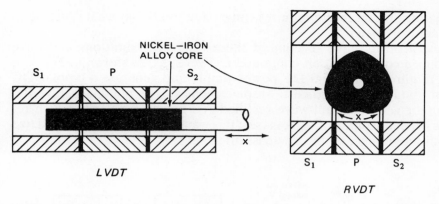

Fig 3.3 Typical construction of an LVDT and RVDT

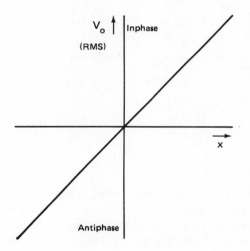

Fig 3.4 Form of the voltage vs displacement characteristic of differential transformers and inductors

- virtually infinite life
- maintenance free
- linear law
- high sensitivity
- relatively inexpensive.

These properties make these transducers appear to be virtually ideal for displacement measurement. However, one must also take into consideration the fact that the useful travel is extremely limited with rotary types, that a special AC reference supply is required, and that the output signal is alternating and therefore requires conversion to DC before it is usable in most control systems.

Manufacturers have recognised these last two limitations and have introduced completely self-contained DC versions, as shown in Fig. 3.5; these require an external DC power supply and generate a bipolar DC output voltage proportional to displacement x. The phase-sensitive de-modulator generates full-wave rectified AC having a polarity which re-verses when the phase of V_o reverses; this waveform is then smoothed to a DC output voltage, by the low-pass filter.

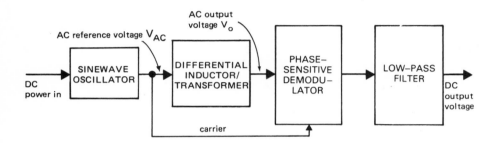

Fig 3.5   Block diagram of a DC-in and DC-out differential inductor or transformer

3.2.3   Capacitive displacement transducers

In recent years, manufacturers have introduced a small number of devices which can be regarded as the electrostatic equivalent of the electromagne-tic differential inductor and which possess comparable advantages. These capacitive transducers are available in both rectilinear and rotary versions and are capable of very fine resolution: for example, rectilinear types have been made which can detect displacements in the μm range. In addition, they can operate in highly hostile environments, including both extremes of ambient temperature.

The alternative principles which can be employed for electrostatic detection of displacement are as follows.

(a)   The introduction of a body into an electrostatic field, resulting in a change in dielectric constant.

(b)   The alteration in the separation distances between capacitance plates.

(c)   The alteration in the effective area of capacitance plates.

In each case, the variation in capacitance can be made to be proportional to the displacement of the input medium. To sense this variation, the capacitor (or differential capacitors) would typically be connected to form the arms of an AC Wheatstone bridge, so that an oscillator and means to convert the bridge AC output voltage to an equivalent DC voltage would normally be required, in order to establish a viable measuring system. This was also the case with the electromagnetic transducers of Section 3.2.2.

3.2.4 Strain gauge displacement transducers

Small displacements may be measured by causing the body being displaced to stretch concurrently, through a suitable linkage arrangement, the element(s) of a strain gauge. (Note that strain gauges are the subject of Chapter 4). The strain gauge circuit generates an output voltage which may be DC or AC, depending upon the type of supply voltage applied, and which is a measure of the elongation of the gauge element(s), and therefore of the displacement applied to the moving input member of the transducer.

Displacement transducers of this type typically have input excursions ranging from about 5 mm to 100 mm, with output excursions in the region of several tens of mV and linearity errors of about ±0.1 mm. Maximum input force required would be in the vicinity of several hundreds of grammes.force.

3.2.5 Synchros

Synchros are very small single-phase rotary transformers, which evolved from now obsolete ranges of transducers known as Selsyns and Magslips. Synchros are supplied in cases with nominal outside diameters ranging from 0.8 inch (size 08) to 2.3 inches (size 23), with sizes 11, 15 and 18 being the most commonplace. Usually, they are manufactured to military specifications, although some commercial versions are available.

Synchros have a cylindrical outer stator and a concentric inner rotor. The stator resembles a miniature slotted and laminated three-phase induction motor stator and carries three star-connected distributed windings. The rotor is laminated, skewed, and may be cylindrical, slotted, and carry one or more distributed windings, or it may have an H-shaped cross section and carry a single concentrated winding.

Normally, electrical connection between the case terminals and the rotor winding(s) is made by means of precious metal brushes running on slip rings mounted on the rotor shaft. However, some manufacturers make a few brushless models, which employ a set of rotating transformers to couple magnetically between the terminals and the rotor winding(s). Typically, each transformer would have a turns ratio of 1:1 and the cylindrical secondary would be mounted coaxially on the rotor shaft and would be surrounded by the cylindrical primary, which would be part of the (stationary) stator assembly: the magnetic coupling would be constant and independent of the shaft angular position.

Synchros are seldom operated singly: usually, they are connected in 'chains' of two or more different types of synchro device. When the ultimate output from a synchro chain is a mechanical displacement, the

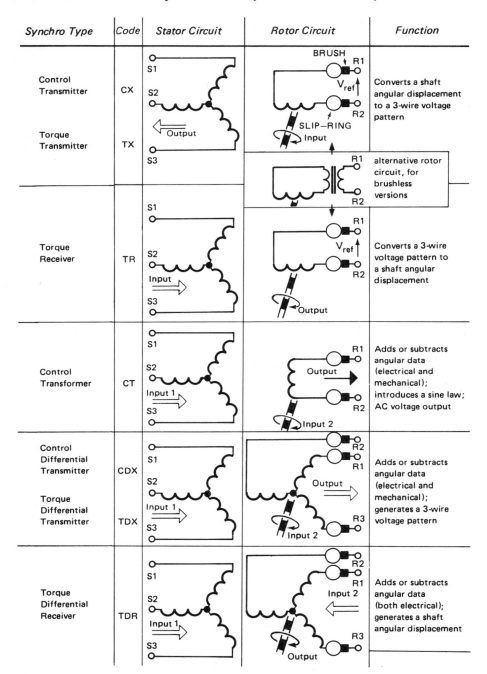

Synchro Type	Code	Stator Circuit	Rotor Circuit	Function
Control Transmitter / Torque Transmitter	CX / TX	S1, S2, S3 Output	BRUSH R1, Vref, R2, SLIP–RING Input. alternative rotor circuit, for brushless versions R1, R2	Converts a shaft angular displacement to a 3-wire voltage pattern
Torque Receiver	TR	S1, S2, S3 Input	R1, Vref, R2, Output	Converts a 3-wire voltage pattern to a shaft angular displacement
Control Transformer	CT	S1, S2, Input 1, S3	R1, Output, R2, Input 2	Adds or subtracts angular data (electrical and mechanical); introduces a sine law; AC voltage output
Control Differential Transmitter / Torque Differential Transmitter	CDX / TDX	S1, S2, Input 1, S3	R2, R1, Output, R3, Input 2	Adds or subtracts angular data (electrical and mechanical); generates a 3-wire voltage pattern
Torque Differential Receiver	TDR	S1, S2, Input 1, S3	R2, R1, Input 2, R3, Output	Adds or subtracts angular data (both electrical); generates a shaft angular displacement

Fig 3.6 Electrical stator and rotor circuits for all synchro types

synchros are said to be 'torque synchros', which are represented by the letter T as the first of a two- or three-letter code. When the ultimate output from a synchro chain is an AC voltage, the synchros are said to be 'control synchros', which are represented by the letter C as the first letter in the code. Torque synchros are normally used for remote signalling of angular data, such as might be required in ships, aircraft, factories, railways, etc., although this does not preclude their use in feedback control systems: for example, the mechanical output displacement could be used to actuate a hydraulic servovalve (see Section 9.2.1) in an electrohydraulic servo-system. Control synchros are used specifically as reference and feedback transducers in feedback control systems.

Figure 3.6 shows the electrical stator and rotor circuits for all of the alternative synchro types, together with the function of each. Manufacturers' data sheets should be complied with rigorously when selecting combinations of synchros for the construction of a synchro chain.

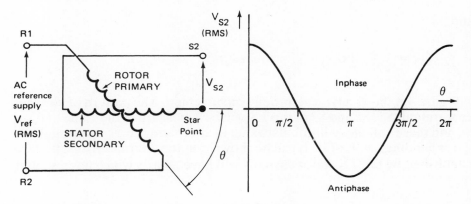

Fig 3.7   Principle of the rotating single-phase transformer

In the case of a synchro transmitter, the rotor winding and any one stator winding behave as a single-phase transformer with variable magnetic coupling, as shown in Fig. 3.7. If the alternating rotor flux is resolved into two components, one in the direction of the axis of the stator winding and the other perpendicular to it, then it is the former component which will be responsible for the generation of the AC EMF in the secondary winding. Thus, in terms of RMS values and ignoring losses

$$V_{S2} = \frac{1}{k} V_{ref} \cos\theta$$

where k is the turns ratio and θ is the relative angular displacement of the rotor. It follows that, for instantaneous values

$$v_{S2} = V_{S2_m} \sin\omega_c t = \frac{1}{k} V_{ref_m} \cos\theta \sin\omega_c t$$

where ω_c is the frequency, in rad/s, of the reference supply and subscript m denotes maximum (peak) value. When θ lies between $\pi/2$ and $3\pi/2$, the phase of V_{S2} will be the reverse of that which exists when θ lies between $-\pi/2$ and $\pi/2$.

Thus, the RMS value of V_{S2}, taken in conjunction with the phase of V_{S2} relative to V_{ref}, is a measure of the value of θ; however, the data would be ambiguous because, for any specific magnitude and phase of V_{S2}, there are two alternative corresponding values for θ, except when V_{S2} is at a positive or negative maximum. Fortunately, when the voltages (V_{S1} and V_{S3}) at the other two stator terminals are also taken into consideration, a specific combination of the three voltages will represent unambiguously a single value for θ. Note that

$$v_{S1} = \frac{1}{k} V_{ref_m} \cos(\theta + 2\pi/3) \sin \omega_c t$$

and

$$v_{S3} = \frac{1}{k} V_{ref_m} \cos(\theta - 2\pi/3) \sin \omega_c t$$

Figure 3.8 shows a typical torque synchro chain. The pattern of voltages on terminals S1, S2 and S3 of the TX will be a measure of the value of θ_i set in, and these voltages will circulate a set of alternating currents through the stator windings of the TR. It can be shown that the alternating flux pattern established by the TR stator currents will have an axis which has a spatial

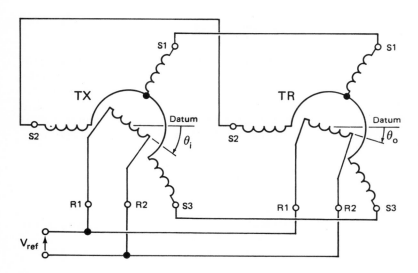

Fig 3.8   Circuit for a TX-TR torque synchro pair

alignment displaced by angle θ_i from datum. Whenever θ_o is different in value from θ_i, the TR rotor flux will react with the stator flux to produce motoring action and the torque generated will drive the TR rotor to align the rotor winding axis with the axis of the stator flux. Thus, in the steady state, θ_o will be a replica of θ_i and the shaft of the TR rotor will duplicate variations in displacement of the shaft of the TX rotor. The system is therefore useful for signalling angular data; the effect of the length of the transmission path on the accuracy of angular data transmission is not of primary significance.

In practice, the TX and the TR will tend to 'fight' one another for dominance: this effect can be resolved by mismatching the stator winding impedances of the two machines and by designing high friction levels into

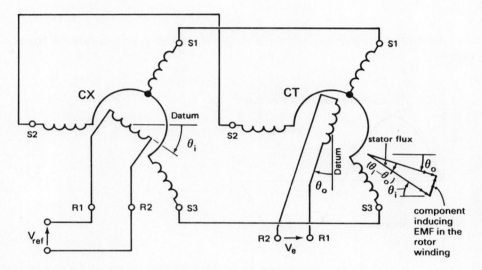

Fig 3.9 Circuit for a CX-CT control synchro pair

the TX rotor shaft drive. In addition, some torque receivers use a rather different construction from that described here: these are known as *torque indicating receivers* and are incapable of generating transmitter action. In such devices, all of the windings are static, with the star-connected windings on a cylindrical outer stator and the fourth winding on a concentric, cylindrical, inner stator. The rotor element is now part of a thin cylindrical metal shell which is free to rotate between the two stators. The torque developed is significantly less than that with a conventional *torque receiver*.

Figure 3.9 shows a typical control synchro chain. Similarly to the torque synchro chain, there will be an alternating stator flux pattern established in the CT, with its axis aligned with a displacement θ_i measured from the direction of the S2 axis. By transformer action, this alternating flux will induce an alternating EMF in the CT rotor winding and the magnitude of this EMF will depend upon the relative angular alignment of the CT rotor winding with the axis of the CT stator flux. By measuring the CT rotor displacement from a datum normal to the direction of the S2 axis, the inherent cosine relationship is converted to a sine relationship yielding

$$V_e = \frac{1}{k'} V_{ref} \sin(\theta_i - \theta_o)$$

for RMS values, and

$$v_e = \frac{1}{k'} V_{ref_m} \sin(\theta_i - \theta_o) \sin\omega_c t$$

for instantaneous values, where k' represents the combination of turns ratios.

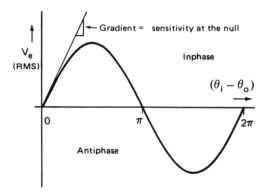

Fig 3.10   Calibration curve for the synchro pair of Fig 3.9

The calibration curve for the CX-CT combination is shown in Fig. 3.10. The value of V_{ref} has been standardised to two alternatives: 26 V and 115 V RMS. A '26 V' synchro pair will yield a gradient of 22.5 V per radian at the origin of this characteristic, so that the maximum value of V_e will be 22.5 V RMS; a '115 V' synchro pair will yield a gradient of 57.3 V per radian (1 V per degree) at the origin, corresponding to a maximum value for V_e of 57.3 V RMS.

This synchro pair would normally be used in a closed loop control system, with the CX serving as the reference transducer and the CT performing the dual roles of feedback transducer and error detector. Figure 3.11 shows a typical servosystem using control synchros. The sine relationship inherent in the V_e vs $(\theta_i - \theta_o)$ characteristic means that such a system will be highly nonlinear in its behaviour, except in the vicinity of the

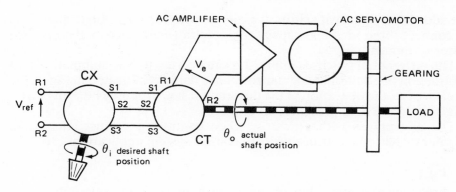

Fig 3.11 Control synchro pair in a closed loop configuration

true null ($\theta_i - \theta_o = 0$) and the false null ($\theta_i - \theta_o = \pi$). The motor will be connected to drive in a direction to increase θ_o whenever V_e is inphase, and to decrease θ_o when V_e is antiphase, with the result that the steady state will always correspond to the vicinity of $\theta_i - \theta_o = 0$, so that $\theta_o \cong \theta_i$, as required. $\theta_i - \theta_o = 0$ then represents a stable (true) null, whilst $\theta_i - \theta_o = \pi$ is an unstable (false) null.

The waveform of V_e will be represented by the expression, for the instantaneous value

$$v_e = \frac{1}{k'} V_{ref_m} \sin(\theta_i - \theta_o) \sin \omega_c t$$

which implies a sinusoidal carrier

$$\frac{1}{k'} V_{ref_m} \sin \omega_c t$$

being amplitude modulated by the instantaneous value of $\sin(\theta_i - \theta_o)$.

The value of the synchro supply frequency has largely been standardised on two alternatives: 60 Hz and 400 Hz. However, other supply frequencies (such as 50 Hz and 1100 Hz) are occasionally used. A higher frequency is preferable because, when the synchros are in motion, V_e also contains a parasitic component arising from generator action: this component is proportionally smaller for higher supply frequencies.

The accuracy obtainable with a synchro chain is in the order of 10 minutes of arc per synchro, which is equivalent to 0.046% of a full revolution. In practice, V_e will be subject to small parasitic phase shifts and will contain small components at frequency harmonics of ω_c, due to the effects of magnetising current, iron losses, winding impedances, and mechanical manufacturing tolerances. These phase shifts can be corrected, using power factor correction techniques employing, for example, the addition of small capacitors at strategic points: a common technique is to connect three identical capacitors in delta (mesh) across the synchro stator terminals.

The differential synchros in Fig. 3.6 can be used for the summation and subtraction of additional angular reference data, as indicated in Fig. 3.12. In each case, the operation of the differential synchros may be deduced by referring to the data presented in Fig. 3.6.

The advantages with synchros, when contrasted against alternative angular displacement transducers in control system applications, are:

- rugged
- breakaway torque is low or, with brushless versions, very low
- stepless static characteristic
- accuracy is relatively insensitive to loading
- high sensitivity
- 360° of useful rotation, with no discontinuity in the static characteristic
- long life
- maintenance is low or, with brushless versions, very low
- moderately expensive.

The principal disadvantages are:

- highly nonlinear static characteristic, except near the null
- require a special AC reference supply for the CX rotor
- error voltage is AC and may require subsequent conversion to DC
- error voltage is subject to parasitic phase shifts
- not available in rectilinear form.

3.2.6 Resolvers

Resolvers operate using the same principles as conventional synchros, but in most cases are constructed with the stator and rotor each having two, electrically isolated, windings distributed to resemble a two-phase type of construction. They can be used for resolving data from rectangular (cartesian) to polar co-ordinate form and vice-versa, and this type of application explains the name given to these devices. Resolvers can also be used for co-ordinate rotation and in place of the CX, CDX, and CT in data transmission and control chains. Figure 3.13 illustrates these various applications.

3.2.7 The Inductosyn

The *Inductosyn* principle is based upon that of a resolver transformer and the device is manufactured in both rectilinear and rotary versions. The

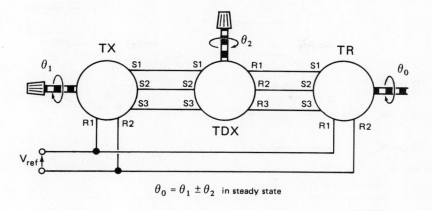

$$\theta_0 = \theta_1 \pm \theta_2 \quad \text{in steady state}$$

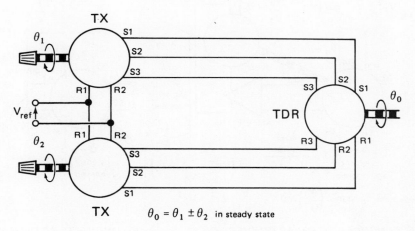

$$\theta_0 = \theta_1 \pm \theta_2 \quad \text{in steady state}$$

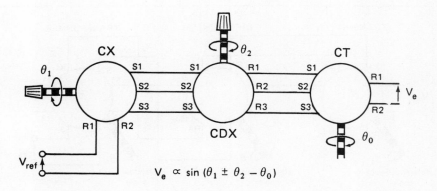

$$V_e \propto \sin(\theta_1 \pm \theta_2 - \theta_0)$$

Fig 3.12   Application of differential synchros for the addition and subtraction of angular data

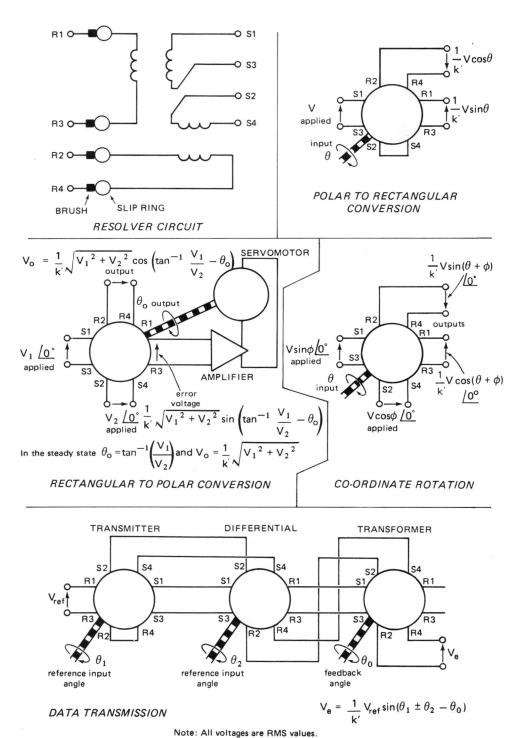

Fig 3.13 Resolver circuit and applications of resolvers

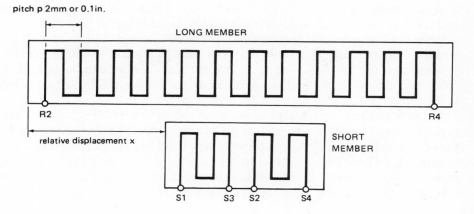

pitch p 2mm or 0.1in.

LONG MEMBER

R2

R4

relative displacement x

SHORT
MEMBER

S1   S3   S2   S4

Note: The windings shown are symbolic representations and are not to scale.

Fig 3.14 Symbolic representation of Inductosyn windings

windings are printed in copper, bonded to a glass or metal substrate, and assume a 'hairpin' form. In the rectilinear version, a short member carries two short hairpin windings, adjacent to one another and separated by three-quarters of a winding pitch, as shown symbolically in Fig. 3.14; in addition, a long member carries a single, continuous, hairpin winding having the same pitch as the windings on the short member. In practice, the short member sits over the long member but is just separated from it. The long members are made typically in 25 cm lengths, which may be mounted together end-on, whilst the short member is typically 10 cm long.

When the short member is excited with a pair of voltages similar to those which would be generated by a resolver transmitter, the Inductosyn behaves as a rectilinear resolver transformer and the winding on the long member generates an AC output voltage with an RMS value given by an expression of the form

$$\frac{1}{k'} V_{ref} \sin(\theta_i - \theta_o)$$

where $\theta_o = \tan^{-1}(x/p)$. The excitation voltages can have any frequency from 200 Hz to 100 kHz, with 400 Hz being typical, and normally they are synthesised using the type of circuitry described in Section 3.3.3. The Inductosyn alone cannot unambiguously interpret axial displacements in excess of $\pm p/2$, so that it must be used in conjunction with other transducers, as discussed in Section 3.2.10. Accuracies in the order of ± 2.5 μm are claimed for the rectilinear types.

Rotary Inductosyns employ two flat circular members, the pair having alternative diameters of 7.5, 10.5, and 15 cm. The hairpin windings in this instance are orientated with their long sides radial from the disc centre and occupy the full 360°: the disc carrying the two windings is printed with the windings divided up into a multiplicity of sectors, whilst the winding on the other disc is continuous. The accuracy of these devices can be as high as ± 0.5 second of arc, which is equivalent to $\pm 3.9 \times 10^{-5}\%$ of one full revolution.

Because of their high degree of accuracy, and even better resolution and repeatability, Inductosyns find application in precision machine tools, automatic inspection systems, theodolites, tracking aerials, inertial navigators, etc. Their life is almost infinite, reliability is very high, and the maintenance required is negligible. Other advantages and disadvantages will be similar to those which apply to synchros.

3.2.8   Shaft encoders

Shaft encoders (which are sometimes called *digitisers*) are digital rotary displacement transducers which are manufactured in electromechanical, optical, and magnetic versions. Shaft encoders may be of the 'absolute' variety, which generate parallel digital data, or the 'incremental' variety, which generate serial digital data typically in the form of pulse trains. Figure 3.15 shows a 4-bit absolute electromechanical encoder, in which the disc is rotated by an input shaft and the brushgear is attached to a surrounding case.

The disc has a metallic (typically, gold-plated copper) pattern printed onto it, using precision printed circuit techniques, and the pattern represents angular data in some form of binary code: normally, metal corresponds to binary 1 and insulation corresponds to binary 0. A DC voltage is

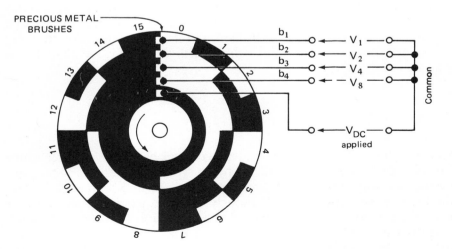

Fig 3.15   Symbolic representation of an absolute shaft encoder

applied to the metallic pattern, by means of a brush running on a printed concentric slip ring. Thus, the radial line of brushes detects the presence or absence of the DC voltage and, in so doing, the set of output voltages indicates the current angular position of the input shaft, in terms of the binary code inherent in the metallic pattern.

The least significant track is always placed outermost, in order to maximise the resolution, which is limited by the dimensions of the brush contact surface and by the diameter of the disc. In order to keep the diameter of the case to reasonable proportions, single disc shaft encoders are not normally manufactured with a word length greater than 10 bits, corresponding to a resolution of 1 part in 1024, which is marginally better than 0.1% of a full revolution. By gearing together two discs, word lengths corresponding to the order of 1 part in 10^5 (0.001%) are achievable in practice, being limited by the precision of the gearing. The combination of two discs to generate composite output data represents a digital application of the 'coarse-fine' technique discussed in Section 3.2.10, with the low-speed disc (which is mounted directly on the input shaft) providing the coarse channel data: in other words, the more significant data byte.

If the binary code is chosen such that the same numerical value is assigned consistently to a particular bit, then the code is said to be 'weighted'. A problem which can exist with all discs having weighted codes is that of 'ambiguity', which arises at every alternate transition in the 'natural binary' coded disc pattern of Fig. 3.15: this phenomenon occurs because of the finite width of the brushes and the manufacturing tolerances on the alignment of the brushes and on the disc production. Thus, when the set of brushes is bridging sectors 1 and 2, 3 and 4, 5 and 6, etc., the pattern of voltages may represent a value completely different from the values represented by the bridged sectors. For example, with the disc in the position shown in Fig. 3.15, the voltages could represent any four-bit pattern of 1s and 0s, depending upon which brush is in contact with metal and which brush with insulation. The solution found to this problem is to allocate two staggered brushes to every binary channel except the least significant; the alternative techniques using this principle have been given such names as 'V-Scan', 'U-Scan', '2-P-Scan', 'V-Disc', etc. The two brushes in any particular channel are gated, using suitable logic, by the state of the output of either the next less significant channel or the least significant channel, depending upon the technique adopted. The gating is such that no brush, except the least significant, can be addressed whenever it is in the vicinity of a transition: the output data generated by the logic are unambiguous, except during the nanosecond duration switching process. Most commercially available encoders of this type incorporate the logic hardware within the case of the encoder.

Other types of binary code which are used quite frequently in encoders fall into the class of 'unweighted' codes, which description applies to Gray, cyclic progressive, and reflective codes. These have the distinction that, at every transition, there is a change in state in only one output channel. Table 3.1 shows the difference between natural binary and Gray code, for

Table 3.1 Comparison of natural binary and Gray codes

Decimal value	Natural binary code				Gray code			
	b_4	b_3	b_2	b_1	b_4	b_3	b_2	b_1
0	0	0	0	0	0	0	0	0
1	0	0	0	1	0	0	0	1
2	0	0	1	0	0	0	1	1
3	0	0	1	1	0	0	1	0
4	0	1	0	0	0	1	1	0
5	0	1	0	1	0	1	1	1
6	0	1	1	0	0	1	0	1
7	0	1	1	1	0	1	0	0
8	1	0	0	0	1	1	0	0
9	1	0	0	1	1	1	0	1
10	1	0	1	0	1	1	1	1
11	1	0	1	1	1	1	1	0
12	1	1	0	0	1	0	1	0
13	1	1	0	1	1	0	1	1
14	1	1	1	0	1	0	0	1
15	1	1	1	1	1	0	0	0

Note b_4—most significant bit b_1—least significant bit

the equivalent decimal values ranging from 0 to 15. Inspection of the Gray code shows that, at every transition, either bit b_1 or b_2 or b_3 or b_4 changes state: this means that, irrespective of the width of each brush and the precision of its alignment and of the disc pattern (provided that the combination does not exceed the width of one least significant bit in the pattern), the output data can only represent either the numerical value just being left behind or the numerical value just being approached. This property is true for all codes in this particular class, as is the characteristic that any particular bit has no consistent numerical value, which property is the meaning of the term 'unweighted'.

In Section 12.3.1 it is shown that, in digital controllers, the processing of data should be undertaken using a weighted code, because of the relative simplicity of the logic required to implement the arithmetic. It follows that, whenever a shaft encoder generating an unweighted code is to be used, the output data must be converted to a weighted format before they can be processed by the controller. The user of the encoder will need to provide the necessary conversion logic.

All of the encoders described so far will require additional logic in some form, either integrated into the encoder case or added externally to the encoder. In every instance, erroneous output data of nanosecond duration will be generated whenever the logic changes state and the controller must be designed so that these transients are insignificant.

All of the encoders described so far have also been of the 'absolute' variety: this implies that the output data format is parallel and that, immediately power is restored following an interruption in supply, the output data word will be restored correctly. An alternative category of encoders is referred to as 'incremental', in which the output signal takes the form of either a pulse train, a squarewave, or a sinewave; the set of output signals would typically take one of the following forms:

	Output A data	*Output B data*
Form 1:	Increments in displacement irrespective of direction	Current direction of motion
Form 2:	Clockwise displacement increments	Counterclockwise displacement increments

Incremental encoders are manufactured using the same techniques as are used for absolute encoders, with the principal difference being that far fewer tracks are required for the incremental types, so that the case diameters tend to be significantly smaller. Figure 3.16 shows a typical disc

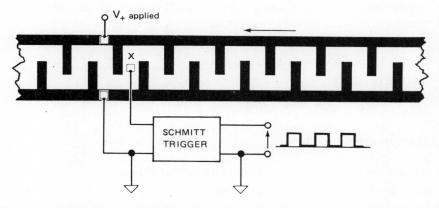

Fig 3.16 Symbolic representation of the track of an incremental shaft encoder

pattern for generating a squarewave representing increments in displacement but not indicating direction. In this case, the Schmitt trigger changes state whenever brush x senses V_+ or 0 V, but it is unaffected by an open circuit input.

With all types of incremental encoder, the output signals must usually be processed by a reversible counter before the data can be used by the controller. The effect of the counter is to convert the serial data generated by the encoder into the parallel data format required by the controller. The principal problems with such a technique are that the count is lost whenever the power supply is interrupted and that the counter may respond to noise spikes on its input lines. When the count is lost, the counter must be zeroed electrically and the encoder drive must simultaneously be zeroed mechanically.

The manufacturing techniques described so far have involved an electromechanical type of construction. A commonly used alternative, which is significantly more expensive, involves optical techniques with which the encoder disc is made from glass and the pattern is printed to form alternate clear and opaque areas. The brush system is now supplanted by a system of photocells which are excited by a light source and lens system. A few manufacturers have developed an electromagnetic alternative, based on comparable principles, in which the disc is now metallic and presents to a set of magnetic sensors a sequence of alternate areas of high and low magnetic reluctance.

The merits of the various shaft encoder types, which are available commercially for measuring only angular displacement, are listed below.

Absolute shaft encoders

- data restored when power supply is restored
- require anti-ambiguity logic, which usually is integral, or code conversion logic, which usually is external
- noise insensitive
- usually coded for positive data values only
- relatively bulky
- require a relatively large number of output connections

Incremental shaft encoders

- require an external counter, in most applications
- data lost when power is interrupted
- noise sensitive
- relatively compact
- require a small number of output connections

Electromechanical shaft encoders

- low maximum speed
- limited life
- poorer reliability and maintainability
- relatively high breakaway torque
- output current must be very small (μA) to avoid brush arcing, which will destroy the disc pattern
- limited maximum resolution per disc
- relatively lower cost

Optical and electromagnetic shaft encoders

- high maximum speed
- almost unlimited life
- good reliability and maintainability
- low breakaway torque
- disc cannot be damaged due to output loading
- greater maximum resolution per disc (optical types)
- relatively higher cost

Compared with other displacement transducers, encoders of comparable resolution are very expensive and this disadvantage has tended to preclude their widespread use; this is despite the obvious merit associated with the digital nature of the output data. In addition, rectilinear versions are not commonplace, except in the form of *diffraction gratings*, which are described next.

3.2.9 Diffraction gratings

These form the basis of transducers which employ an optical principle of operation and which are directly competitive with the Inductosyn. Like the Inductosyn, gratings are manufactured in both rectilinear and rotary versions; unlike the Inductosyn, gratings do not require to be incorporated into a coarse-fine measuring system in order to create a viable transducing arrangement.

The rectilinear version of the transducer involves a long grating, which is usually the moving element, and a short grating, which is usually the stationary element. The short grating, which is typically about 2 cm square, is made from glass and has ruled upon it a series of parallel straight opaque lines, between which will remain a series of parallel straight clear lines. On a 'coarse' grating, the lines would typically be pitched 10 or 40 to the millimetre whilst, on a 'fine' grating, they might be 400 to the millimetre.

The long grating would typically be 25 cm long by 2 cm wide and further gratings would be abutted end-on whenever travels in excess of 25 cm are required. The long grating is made from steel, for light-reflecting versions, and from glass, for light-transmitting versions; this grating is also ruled with parallel lines having the same pitch as the lines on the short grating.

Figure 3.17 shows how the two gratings would be mounted relative to each other, with the optical system shown for a light-transmitting version. With the light-reflecting type, the light source and photocells would both be mounted above the short grating. Note that the lines on the short grating are inclined, relative to those on the long grating, and this results in a set of light and dark bands, each typically about 0.5 cm wide, running lengthwise (that is, perpendicular to the lines on the long grating).

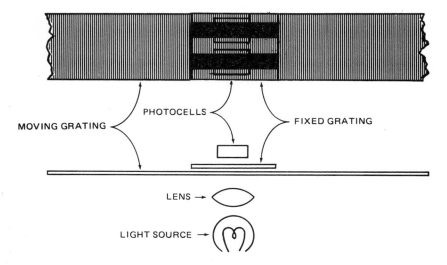

Fig 3.17 Symbolic representation of the construction of a diffraction grating

When the long grating is displaced by one half of one line pitch, the bands move across the grating, with a light band now replacing a dark band and vice-versa: the direction of movement of the bands reverses whenever the direction of motion of the grating reverses. The photocells, by sensing the presence of a light or dark band, with the associated circuitry can sense a displacement typically of one quarter of one line pitch: that is, 0.6 μm in the case of fine gratings. Special versions have been developed which can resolve down to one tenth of one line pitch. The photocells will average the light distribution across many lines, so that imperfections in individual lines will be of no significance.

This measuring technique, which here produces optical magnifications of up to (say) 4000:1, uses a phenomenon known as 'Moiré fringes'. In the case of fine gratings, optical 'interference' results in enhancement of the light and dark bands produced by these fringes, so that the discrimination by the photocells is improved.

The fixed and moving gratings are mounted in close proximity without being in physical contact, so that frictional forces are zero. The output signal from the photocell circuits will usually be a pulse train, together with a second signal indicating the direction of motion; alternatively, some commercial products generate sinusoidal output signals having a phase relationship which is a measure of the grating displacement. In the case of the digital output versions, the outputs will typically be applied to a reversible digital counter, with the output count representing the nett displacement of the moving grating occurring since the counter was last zeroed: the data will be lost whenever the power supply is interrupted and can also be corrupted by impulsive noise induced in the interconnections.

Angular diffraction gratings are manufactured using the same principles, with the moving grating being a circular disc and the fixed grating a sector of a circle having the same diameter: the lines are radial, nominally, but slightly inclined to the radius in the case of the fixed grating, and are ruled within an annular band.

Life, reliability, and maintainability of diffraction grating transducers will be of a high order, due to the complete absence of electrical connections in motion and of parts subjected to wear.

3.2.10 Coarse-fine (dual-speed) measuring systems

The accuracy and resolution of a measuring system may be extended beyond that obtainable with a single feedback transducer, if a second and even a third feedback transducer are added to the system. Figure 3.18 shows a displacement measuring system in which the reference and feedback transducers are duplicated and geared to their counterparts.

In the vicinity of the null $(\theta_i - \theta_o = 0)$, the fine channel, in which the

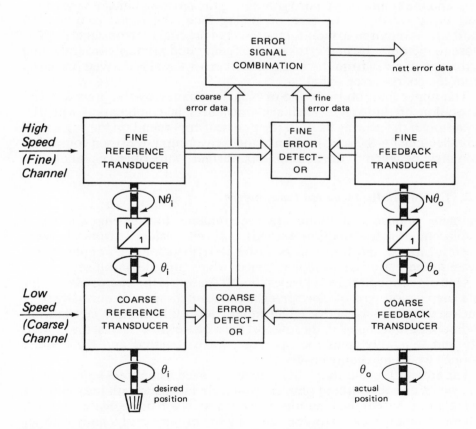

Fig 3.18 Block diagram of a coarse-fine (two-speed) displacement measuring system

output data change at N times the rate at which the output data in the coarse channel change, generates error data having N times the sensitivity (with respect to the error $\theta_i - \theta_o$) of that of the error data generated by the coarse channel. Thus, if the two feedback transducers are identical, the improvement in measurement accuracy and resolution should be N-fold: note, however, that often there is no specific requirement for these two transducers to be identical.

The two reference transducers may not be geared together, if it is desired to be able to independently set each by hand, with the setting of the fine transducer acting as a vernier augmentation of the setting of the coarse transducer. Alternatively, and especially if the requirement is for digital reference data, it may be possible to generate both components of reference data using a single data generation system.

The error signal combination hardware will have to generate nett error data from the combining of coarse and fine channel error data. In the case of digital data, this may simply be an instance of combining end-on the two error data words, with the fine error word generating the less-significant byte and the coarse error word generating the more-significant byte of the nett error word. In the case of analog data, the signal combination hardware is required to monitor the value of the coarse error signal so that, close to the null, the nett error signal is generated primarily from the fine error signal; away from the null, the nett error signal is derived primarily from the coarse error signal.

The upper limit on the value of N is determined by the precision and resolution of the coarse feedback transducer in combination with the precision of the gearing itself. If the practical upper limit on the value of N is inadequate for the specification of the measuring system then consideration must be given to the addition of a further ('superfine') channel.

3.2.11   On-off displacement transducers

In many applications, there are requirements for sensing whether a displacement is less than, or exceeds, a specific value. It would be very wasteful to use the transducers so far described for such applications, because there are many simpler devices which provide on-off sensing.

One of the simplest and cheapest on-off sensors is the *microswitch*. This is a snap action type of electromechanical switch which is normally fully enclosed and may be activated by either a lever or a plunger mechanism. Being electromechanical, the low cost must be weighed against the limited life and reliability, plus the operating force required and the contact bounce which inevitably ensues.

Far greater life and reliability can be achieved using 'proximity switches', which operate without physical contact being made with the switching mechanism, so that the operating force required will be negligible or zero. There are many techniques for detecting the proximity of a body without making physical contact with it, and amongst these would be included:

- the introduction or removal of a magnetic field
- the change in reluctance of a magnetic circuit
- the change in permittivity of the dielectric of a capacitor
- the change in physical dimensions of a capacitor
- the change in back pressure within a nozzle supplied, through a restriction, with compressed air (see Section 9.1.2)
- the change in reflection, refraction, or transmission of a light beam
- the change in reflection, refraction, or transmission of an ultrasonic beam.

The details of the proximity switches which are available commercially would occupy a large catalogue and the user is best referred to suppliers' data books for this type of information. The types of parameter which would be relevant when making a choice for a specific application would include:

- physical size
- limits on physical clearance
- mounting details
- suitability of the enclosure for the plant environment
- operating force requirements
- power supply requirements
- output signal characteristics
- settling time
- possibility of interference with, or by, the operating medium.

3.3   Reference transducers

Whilst it may seem to be illogical to include a discussion on reference transducers in a chapter devoted principally to feedback transducers, the justification for this is based upon the fact that many reference transducers are similar to the displacement transducers described in Section 3.2, especially where there is to be manual setting of the reference data.

The most commonly used devices for setting manually the reference variable are:

- potentiometers
- synchro and resolver transmitters
- rotary switches
- pushbutton, key, and toggle switches.

Where a digital computer is to be used for establishing the reference data, there is often a requirement for the data word generated by the computer to be converted to a form (analog or digital) which is appropriate for the control system, and the reader should refer to Chapter 13 for examples of systems with this requirement.

3.3.1   Potentiometric references

Reference potentiometers are generally required to have an accuracy similar to that associated with servo potentiometers. The existence of a deadspace and/or limits to travel no longer presents a problem, however, and a reasonably high level of friction can often be a definite advantage. Electrical loading errors can be as significant a problem as with feedback potentiometers, so that unloading buffer amplifiers should be incorporated wherever necessary.

There is no requirement for the voltage supply for the reference potentiometer to have a high value, whatever the maximum value of the feedback signal: see Section 12.1 for discussion on this. Theoretically, almost any voltage supply level can be used; in practice, the level should be sufficiently high to maintain high signal/noise ratios but not so high as to generate significant heating of the potentiometer circuit or to render the supply hardware unnecessarily expensive. There is a definite requirement for the supply voltage to be stable, because the stability of the reference transducer sensitivity depends upon this: a possible exception would be a system in which the reference and feedback transducers are to use a common voltage reference supply: the reference and feedback transducer sensitivities would then change by comparable factors, should the supply voltage change, and the system loop gain would also change by a similar factor.

Figure 3.19 shows typical reference networks for position and speed control systems. Notice that:

- for a precise electrical null, it is preferable to relate zero reference to signal common potential
- if V_+ and V_- are not carefully matched in value, preset trimmer potentiometers should be included especially, for voltage balancing purposes – this will not apply to undirectional systems, which only require a unipolar reference
- the input dial preferably should have a linear scale and be calibrated in terms of desired value.

3.3.2   Synchro and resolver transmitters

Where synchros, resolvers, or Inductosyns are to be used as feedback transducers, it is commonplace to use comparable synchros and resolvers as the reference transducers. In these cases, it may be advantageous to build additional friction into the reference shaft bearings, to prevent the

TYPICAL REFERENCE NETWORK FOR A POSITION CONTROL SYSTEM

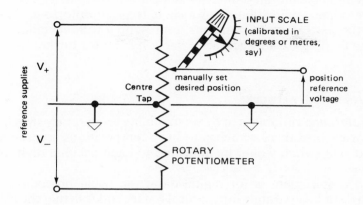

TYPICAL REFERENCE NETWORK FOR A SPEED CONTROL SYSTEM

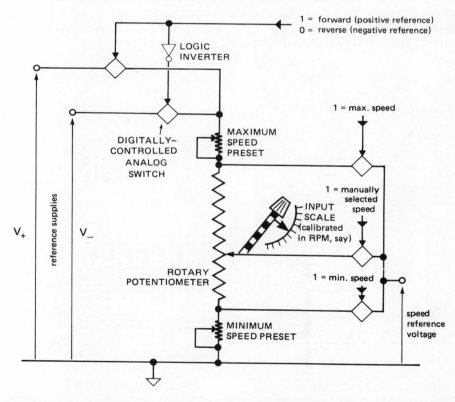

Fig 3.19   Typical reference networks for position and speed control systems

setting from being moved by vibration. The dials would normally carry linear scales and be calibrated in terms of desired position.

There are some applications in which it is preferable to synthesise electrically the pattern of AC voltages, represented reference data, presented to the feedback transducer. See Section 3.3.3 for further information.

3.3.3   Rotary switch networks

Where the reference variable is to be set manually using rotary switches, it is normal for each of these switches to have ten positions (representing 0–9), so that one switch would be allocated to each decimal digit of the desired value.

When the requirement is for digital data, the usual procedure is to encode each decade separately into natural binary code, so that the overall format for the reference data becomes binary-coded-decimal (B C D). Figure 3.20 shows two commonly used alternatives for generating natural binary code for one decade of the B C D word. Some manufacturers will supply the switches with integral logic hardware. Since most digital control systems do not process B C D data, it frequently becomes necessary to convert the reference data to the code of the processor.

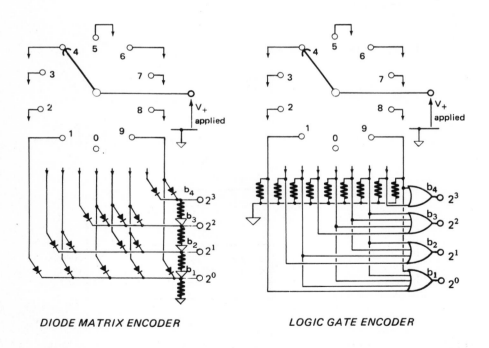

DIODE MATRIX ENCODER   *LOGIC GATE ENCODER*

Fig 3.20   Typical switch circuits for generating natural binary code for one decade of BCD reference data

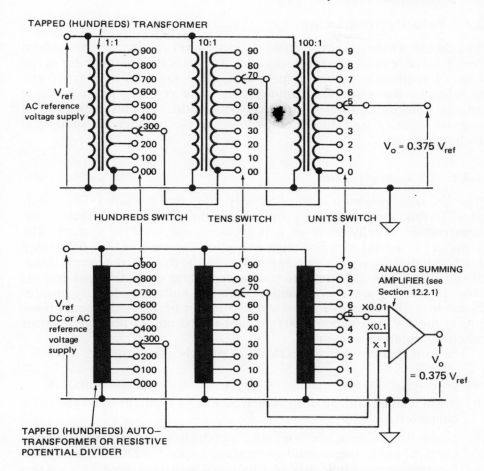

Fig 3.21 Analog switch circuits networks, using tapped transformers and potentiometers, for generating reference signals

When the requirement is for analog data the switches are arranged differently, and two examples are shown in Fig. 3.21.

When the requirement is for bidirectional data to be generated by rotary switch networks, it is necessary to make special provision. One method is to include a sign switch and to cause the selection of (say) a negative sign to complement the digital data (which must now include a sign bit), sign-invert the DC output voltage, or reverse the phase of the AC output voltage, as the case may be. One alternative is to offset the calibration of the reference switches and feedback transducer alike, so that each datum is now represented by (say) the mid-range output data value of the associated network or transducer: thus, for example, nineteen-way switches might now be used for the reference network, with the central setting of each representing zero.

3.4 Velocity transducers

The choice of transducer to measure velocity is very limited, in comparison with the range available for displacement. This is true particularly in the case of rectilinear velocity, for which there are few devices of any significance manufactured commercially. The commonly used analog angular velocity transducers are electromagnetic machines, whilst digital angular velocity transducers employ either electromagnetic or optical techniques.

3.4.1 DC tachogenerator

The *DC tachogenerator* is an accurately calibrated miniature DC generator. To function, it requires a constant excitation and, for this reason, the construction usually involves a permanent magnet field system. The armature is wound with fine wire and, partly because of this, only small load current values (in the mA region) should be drawn from the machine. By so minimising the armature resistance voltage drop, the user ensures that, to a very good approximation, terminal voltage is directly proportional to shaft velocity. Typical sensitivities for permanent magnet types are in the 3 V to 7 V per 1000 rpm range. The output polarity reverses when the rotation is reversed.

The principal problems with this type of tachogenerator are:

- poor reliability and maintainability associated with commutators
- brush contact voltage drop, which imposes a small offset on the static characteristic
- commutator ripple, which manifests as an alternating (non-sinusoidal) parasitic noise component superimposed upon the DC output signal – to some extent, ripple may be attenuated with a low-pass filter but this cannot be completely satisfactory, because the noise frequency varies in proportion to shaft speed and excessive filter time constant could harm the dynamic behaviour of the system for which the tachogenerator is providing the feedback signal
- relatively high breakaway torque
- relatively high polar moment of inertia.

3.4.2 AC drag-cup tachogenerator

The principle of operation of this machine is indicated in Fig. 3.22. The two windings are aligned axially at 90° to one another and are distributed in slots around one or two stators. In some versions, a single outer stator carries both windings; in others, the outer stator carries one winding and an inner stator carries the other. The rotor is usually a hollow cylindrical copper cup driven from one end by the input shaft. A constant AC reference voltage (usually a standard synchro reference) is applied to one

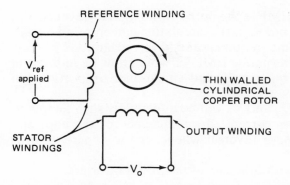

Fig 3.22 Symbolic representation of an AC drag-cup tachogenerator

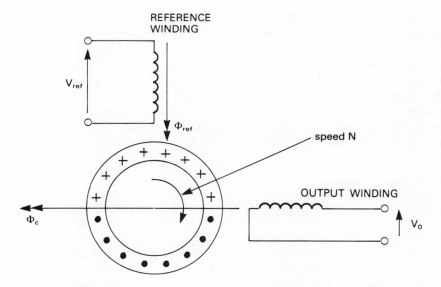

Fig 3.23 Diagrammatic representation of the principle of operation of the AC drag-cup tachogenerator

winding and this establishes a pulsating magnetic flux, with a stationary axis, in which the cup is free to be turned. See Fig. 3.23 for an explanation of the principle of operation. The reference flux Φ_{ref}, in which the cup is turned, is established by the current circulated through the reference winding by the applied voltage V_{ref}. EMFs are generated in the metal of the cup and these will circulate currents in the directions shown. The flux pattern established by these currents can be represented by the flux vector Φ_c.

If V_{ref} were DC and the speed N constant, then the currents circulating in the cup would be constant and DC, so that flux Φ_c would also be constant: it would induce no EMF in the output winding. However, if V_{ref} were to alternate, at a frequency ω_{ref}, the circulating currents and flux Φ_c would also alternate at the same frequency and have a magnitude proportional to shaft speed N: an alternating EMF V_0 would now be induced in the output winding, with a frequency ω_{ref} and a magnitude proportional to speed N.

The phase shift between V_{ref} and V_0 will be affected by such factors as magnetising current, iron loss current, the inductance of the cup, etc. However, there will be a phase reversal whenever the cup rotation is reversed.

The principal advantages with this type of tachogenerator are:

- high reliability, maintainability, and life, due to complete absence of moving electrical contacts
- very low breakaway torque
- very low polar moment of inertia.

The principal disadvantage is that the output voltage is AC (albeit at the fixed reference supply frequency) and this is subject to a parasitic phase shift, the value of which can vary as a function of shaft speed. The output may be converted to DC, using a phase-sensitive demodulator.

When excited with DC, this machine will generate a DC output voltage having a value proportional to the shaft angular acceleration.

3.4.3   AC signal alternator

This machine is a miniature single-phase alternator with a rotating permanent magnet field system. The generator produces a sinusoidal AC output voltage having both amplitude and frequency proportional to the shaft speed. Because of the absence of a reference phase, this machine is unable to indicate the direction of rotation.

Since the frequency of the output is related precisely to the shaft speed, the alternator can be used in digital applications after the sinusoidal output wave has been converted to a suitable squarewave, using a zero-crossing detector: a digital frequency counter can then be used to generate a numerical indication of the shaft speed.

This machine will be characterised by high reliability, maintainability, and life, due to the complete absence of moving electrical contacts; breakaway torque will be low but the polar moment of inertia will be relatively high.

3.4.4   Digital velocity transducers

Any of the digital displacement transducers discussed in Section 3.2 can be used to generate velocity data, if a serial digital output is made available for input to a digital frequency counter. Thus, absolute shaft encoders (using the output from one channel only), incremental shaft encoders, and optical diffraction gratings can all be used as primary sensors for velocity measurement. Additional logic will be required in order to generate an indication of the direction of rotation: this could be complex in the case of absolute encoders.

Dedicated digital velocity transducers are available commercially, and most of these use a castellated disc which causes either a light beam to be interrupted or the reluctance of a magnetic circuit to be stepped. In both cases, the sensing circuit will generate either a squarewave or a pulse train, the frequency of which can be converted, using a digital frequency counter, to a digital word representing the speed of the disc shaft. An alternative technique is to cause the change in magnetic reluctance to modulate the frequency of an oscillator and to convert the modulation frequency to a DC voltage. The electronics may be packaged with the sensor, as a single unit. Some versions will also give an indication of the direction of rotation, by incorporating a duplicate, offset, disc.

With the exception of electromechanical shaft encoders, digital velocity transducers will be characterised by high reliability, maintainability, and life and by low breakaway torque; the polar moment of inertia will depend upon the principle of operation used.

3.4.5   Bridges to measure back-EMF

In Section 8.6.2, the control of DC motors for use in servosystem applications is discussed. In those configurations in which the motor is operated with constant field excitation and the armature is controlled by means of a suitable servo-amplifier, an approximate measure of the shaft speed may be obtained by incorporating the armature circuit into a type of Wheatstone bridge arrangement, as shown in Fig. 3.24.

If the differential amplifier has a gain of K, then its output voltage V_0 is given by

$$V_0 = \frac{K}{2}[E_B + I_A(R_A + R'_A)] - KI_A R'_A \qquad \text{with } K \ll 1 \text{ normally}$$

If the external resistor R'_A is matched in value to the armature resistance R_A of the motor, the expression reduces to $V_0 = KE_B/2$, so that this voltage can be fed back to the signal combination network of the servosystem. Provided that it is justifiable to assume that the back-EMF E_B is proportional solely to the motor shaft speed then it will be valid to use V_0 as a speed feedback signal.

The following factors are relevant to this technique:

● no velocity transducer is required
● R'_A will have a very low resistance value but must have a current rating to match that of the motor armature

- the values of R_A and R_A' will both be affected by self heating and it will be difficult to match them over a wide range of loading

- armature reaction will tend to corrupt the linearity of the E_B vs N characteristic, but to some extent this effect can be compensated for by slightly increasing the value of R_A'

- the use of a differential amplifier assumes that it is impractical to connect either of its two inputs to signal common, which sometimes may not be the case

- R can have a relatively high value, so that the R + R limb need not draw a significant level of current from the servoamplifier

- the accuracy obtainable with this technique cannot approach that provided by a good velocity transducer.

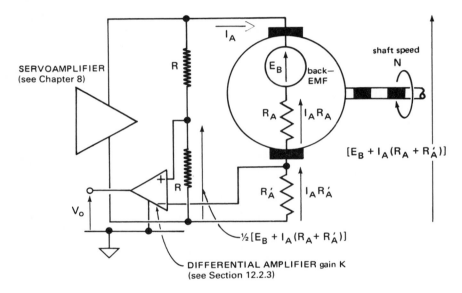

Fig 3.24   Use of a bridge network with an armature-controlled DC motor to generate a speed feedback signal

4

Transducers –
Strain, Force, Torque, Acceleration, Load and Tension

4.1 Introduction

In this chapter, a brief survey of strain gauges and the electrical networks in which they are connected will be undertaken. It is relatively uncommon for strain to be the controlled variable in a control system but at the same time other, more common, controlled variables—such as force, torque, displacement, acceleration, etc.—may be measured by translating them into the strain of a suitable mechanical element and then, in turn, by using one or more gauges to transduce the strain into an electrical feedback signal. Many of the more common applications of gauges as integral components in feedback transducers will then be considered. Finally, a number of alternative techniques for transducing the same types of controlled variable will be presented.

4.2 Strain gauges and measuring bridges

4.2.1   Strain gauges

A strain gauge consists of a wire or foil element mounted on paper or some other low-modulus backing material. When the strain gauge is attached, using adhesive, to the surface being instrumented, it will undergo the same strain variations as the material to which it is bonded. Except in the case where the gauge is attached to a thin foil, the gauge itself will have little influence on the strain variations in the base material.

When the gauge metal is subjected to strain, its electrical resistance changes and this change is detected and used to monitor the cause of the strain. Two factors will contribute to this change in resistance:

- the change in length and cross-sectional area of the element
- the change in the resistivity, of the gauge material, resulting from the strained condition.

In metal gauges, these two factors are roughly equal in significance. The *gauge factor* k is defined by the formula

$$\frac{\Delta R}{R} = k\varepsilon$$

where ΔR is the change in the original resistance value R due to the strain ε. The strain ε is defined by $\varepsilon = \Delta x/x$, where Δx is the extension of the original effective length x of the gauge element, due to the external cause. A typical value for k is 2 and, since ε can have values typically in the range from 10^{-6} to 10^{-3}, it follows that relatively small values of $\Delta R/R$ have to be sensed.

The resistivity of metals is affected by temperature, in addition to strain. Strain gauges often use temperature-compensated elements, in which the elements are constructed using two materials so matched that the temperature-dependent resistance change of one just balances that of the other, over a specified temperature range. Where the gauge element is bonded to a base material, the compensation requirements will depend upon the type of base material used.

Strain gauges are also made using semiconductor materials. These have a higher gauge factor, in the order of 50, and are more temperature-sensitive and more expensive.

4.2.2 Strain gauge bridges

The usual method of detecting the resistance change of a strain gauge is to incorporate it into a Wheatstone bridge, as shown in Fig. 4.1.

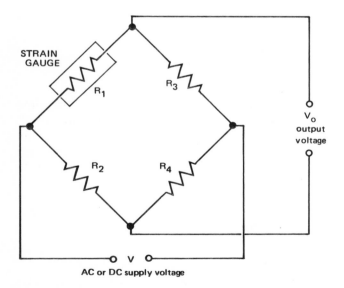

Fig 4.1 Strain gauge in one limb of a bridge network

When the bridge is balanced, $V_o = 0$ and $R_1/R_2 = R_3/R_4$. If R_1 changes, then the change in output voltage is given by $V_o = V/4 \cdot \Delta R_1/R_1 = V/4 \cdot k\varepsilon$, provided that all four resistances are nominally equal. In this example, only the element R_1 is active, whilst the other three resistors are passive and must be highly temperature stable.

When two or four of the bridge elements are active, the bridge can be made sensitive to specific strain conditions. Figure 4.2 is an example in which R_1 and R_2 are used to detect bending strain, in such a way that axial strain effects cancel electrically. Bending strains will, for example, increase R_1 and decrease R_2, and these effects are additive in the resultant change in V_o. Axial strains will, for example, increase R_1 and R_2 equally, and these effects will cancel so as not to influence V_o.

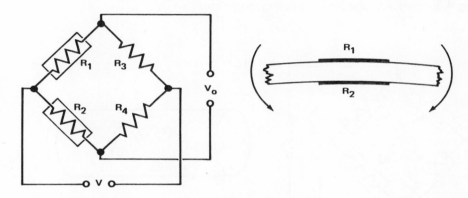

Fig 4.2   Strain gauges instrumenting bending strain

The bridge can be rearranged so as to sense axial strains and ignore bending strains, as shown in Fig. 4.3. To sense torsional strain, for which both axial and bending strains are ignored, all four elements are made active and mounted as shown in Fig. 4.4. The pair of elements R_1 and R_2 will change equally (in resistance) as a result of both bending and axial strains, so that their effects on V_o will cancel to zero. A similar situation will exist for the pair R_3 and R_4. However, torsional strain would typically cause R_1 and R_4 to increase equally and R_2 and R_3 to decrease equally, so that all four effects would be additive in influencing V_o.

Although strain gauges are designed to be sensitive along one nominated axis only, it is inevitable that they should be slightly sensitive to strains along the perpendicular axis. This results from the configuration of the elements, as shown by Fig. 4.5. The wire-type gauge can be up to 5% cross-sensitive whereas the foil-type can be well below 1%.

Some bridge configurations make use of a combination of active and 'dummy' gauges. The dummy gauge is attached to the same base material as the active gauge and mounted in close proximity to it (but aligned along an unstrained axis), so that both gauges (which are electrically identical)

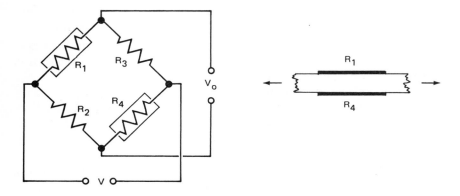

Fig 4.3   Strain gauges instrumenting axial strain

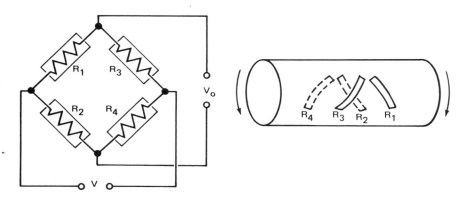

Fig 4.4   Strain gauges instrumenting torsional strain

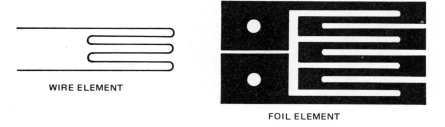

WIRE ELEMENT

FOIL ELEMENT

Fig 4.5   Examples of wire and foil strain gauge elements

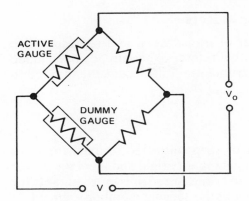

Fig 4.6 Active and dummy strain gauge elements in a bridge network

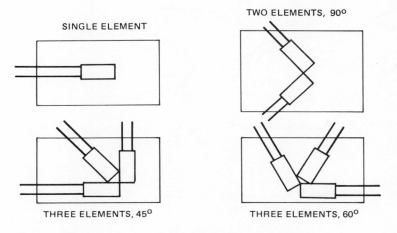

Fig 4.7 Examples of multiple-element strain gauges

will always experience the same temperature. With active and dummy gauges connected in adjacent arms of the bridge, as shown in Fig. 4.6, temperature effects will cancel out and therefore will not influence V_o. When gauges compensated for temperature on steel backing are attached to an aluminium body, for example, the use of a dummy gauge is important.

Strain gauge configurations may be of the single-, two-, or three-element type, as illustrated symbolically in Fig. 4.7. Multiple-element gauges of the foil type are usually referred to as 'rosettes'. Four-element rosettes, with elements aligned along one or two (perpendicular) axes, are also manufactured, for use in full bridge circuits. Multiple-axis rosettes are often used in applications where the direction of the strain is not known beforehand; in transducer applications, the direction is usually specific, so that rosettes would not then be used, as a rule.

4.3   Use of strain gauges in transducers

Strain gauges are used in a number of transducers, and representative applications will be described. Note that reference has already been made to the use of strain gauges for instrumenting displacement, in Section 3.2.4.

4.3.1   Force transducers

The simplest configuration is a bar or strip subjected to axial force, as shown in Fig. 4.8. This type of transducer is very rigid and ideal for tensile forces. When compressive forces are to be measured, it is more suitable for small forces, because of the tendency of the sensing member to buckle.

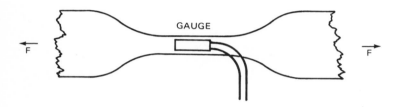

Fig 4.8   Strain gauge instrumenting axial force

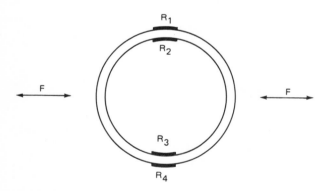

Fig 4.9   Strain gauges mounted on a load ring

To instrument both tensile and compressive forces of comparable magnitude, the sensing member is designed to distort in shape. Figure 4.9, which shows a 'load ring' sensor, is designed for this type of application. The elements will be attached as shown and connected electrically to sense bending stresses, as shown in Fig. 4.10.

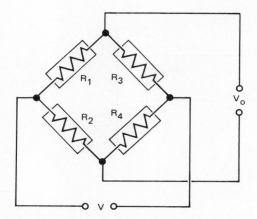

Fig 4.10   Electrical connection of the strain gauge elements in Fig 4.9

4.3.2   Torque transducers

Where torque is to be measured on a stationary body, there are three optional techniques, as illustrated in Fig. 4.11. The choice is largely a matter of convenience and accessibility.

When torque is measured in rotating shafts, the sensing configuration is almost always of the type shown in Fig. 4.11a. It frequently happens that shafts are greatly overdesigned and the strain values to be measured are too small to give good signal-to-noise separation. In such cases, the use of semiconductor strain gauges in half or full bridge configuration is indicated. Where the environment (mainly temperature) precludes this, the shear strain can be intensified as shown in Fig. 4.12. A sleeve is pushed over the shaft and rigidly attached in planes A and B. In plane C, the section of the sleeve is greatly reduced. As a consequence, almost all of the twisting from A to B is concentrated in the reduced section of the sleeve. The strain gauges are located at that point.

Transferring the strain gauge signal from a rotating body to the external environment has been attempted in various ways and some in current use involve:

- using slip rings
- using an RF carrier from a transmitter attachment to the shaft
- using inductive or capacitative coupling and frequency-modulating a carrier.

Each of these methods contributes a certain amount of additional noise to the signal and requires that various pieces of apparatus be attached to the shaft.

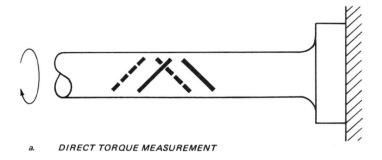

a.   **DIRECT TORQUE MEASUREMENT**

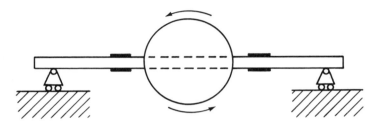

b.   **TORQUE MEASUREMENT BY BENDING A BEAM**

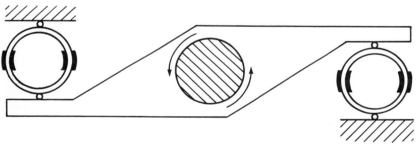

c.   **TORQUE MEASUREMENT BY
       MONITORING REACTION FORCES**

Fig 4.11   Alternative techniques for using strain gauges for instrumenting
torque

By measuring reactions at prime mover or gearbox foundations, the
need for 'on the shaft' torque measurement can often be avoided and the
additional difficulty of getting the signal from the shaft is then avoided.

4.3.3   Acceleration transducers

The problem of measuring acceleration is transformed into the measurement of the force which results when a body of known mass is subjected to

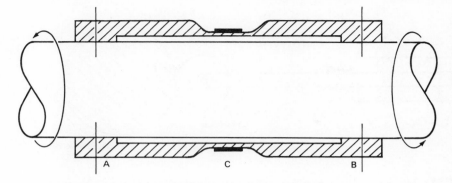

Fig 4.12   Use of a sleeve to intensify torsional strain

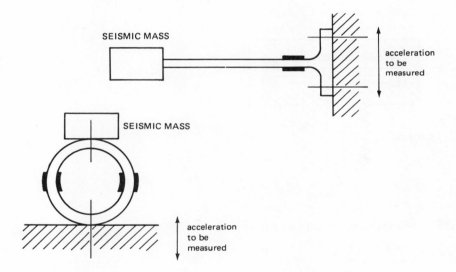

Fig 4.13   Use of seismic masses to convert acceleration into strain

the acceleration to be measured. A body used for this purpose is known as a 'seismic mass' and two possible applications, for the measurement of linear acceleration, are shown in Fig. 4.13.

To measure the angular acceleration of shafts, one may mount accelerometers such as those of Fig. 4.13 on them and measure the tangential acceleration. When this is done, the problem of transmitting the signal to the external environment arises again.

A commonly used expedient that avoids the problem of signal transfer is illustrated in Fig. 4.14. A roller is pressed against the shaft and rolls freely without slip. Angular acceleration shows up as tangential forces which are sensed and give a measure of this acceleration.

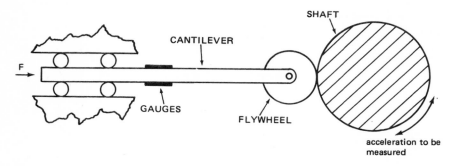

Fig 4.14 Technique for instrumenting angular acceleration of a rotating shaft

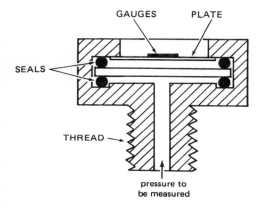

Fig 4.15 Flexing plate pressure sensor

4.3.4 Pressure transducers

The most common type of pressure sensor using strain gauges is the 'flexing plate sensor'. This has a high speed of response and is illustrated in Fig. 4.15.

4.3.5 Load cells

When the weight of a body has to be sensed, the weight can be used to compress a member of the types illustrated in Fig. 4.8 and Fig. 4.9, by physically connecting the scale pan to the member. Alternatively, the weight on the scale pan may be used to pressurise hydraulic fluid, using a cylinder and piston, and the hydraulic pressure so established can then be sensed by a pressure transducer, possibly of the type shown in Fig. 4.15.

4.4 Other transducers for measuring force, torque, acceleration and tension

4.4.1 Force, torque, and acceleration transducers

Force may be measured by firstly causing the force to extend or compress a linear spring of known spring rate, and then instrumenting the spring deflection with a rectilinear displacement transducer.

Torque on a stationary body may be measured by causing the torque to extend or compress a torsional spring of known spring rate, and then instrumenting the spring deflection using an angular displacement transducer.

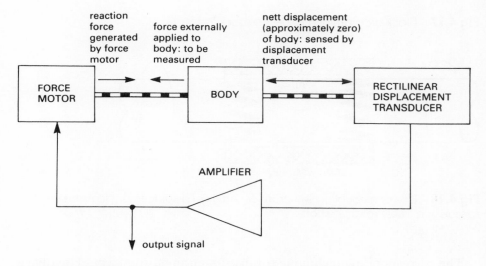

Fig 4.16 Schematic arrangement of a force feedback loop

An alternative arrangement for measuring force is to sense linear displacement of the body to which the force is being applied, using a suitable displacement transducer, to amplify the transducer output signal and to use the amplified signal to excite a force motor (see Section 8.6.4). The force developed by the motor is applied so as to oppose the force being measured, using a suitable mechanical connection, as depicted in Fig. 4.16. The block diagram for the resulting force feedback loop is shown in Fig. 4.17. With the system designed to have a high loop gain, the closed loop action will tend to keep the transducer output signal at a very low level, so that a force balance will have been established. The signal delivered by the amplifier to the force motor will be a measure of the force being fed back and hence will also be a measure of the force which is to be instrumented.

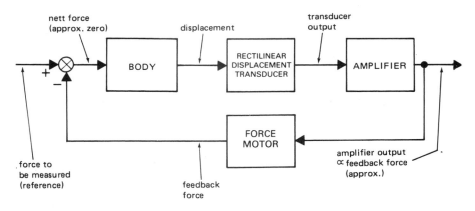

Fig 4.17   Block diagram of a force feedback loop

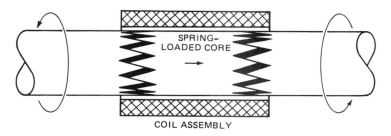

Fig 4.18   Use of a coupling with transduced axial displacement to measure torque in a freely rotating shaft

The corresponding technique for instrumenting torque would involve a rotating body, with the loop being completed by an angular displacement transducer, amplifier and torque motor (also described in Section 8.6.4).

Linear acceleration may be measured using the above techniques, if a seismic mass is subjected to that acceleration and the resulting force is measured. Angular acceleration may similarly be measured, if the resulting torque on the seismic mass is measured.

Angular acceleration may also be measured for a freely rotating body, by coupling to it an *AC drag-cup tachogenerator* of the type described in Section 3.4.2. If such a machine is excited from a constant voltage DC supply, the output voltage generated will now be a DC voltage proportional in magnitude (and representative in polarity) to the shaft angular acceleration.

Torque in a freely rotating shaft may also be measured by incorporating a coupling of the type shown in Fig. 4.18. The central section, which is axially spring-loaded, is displaced axially by an amount proportional to the level of torque being transmitted. If this section is made the (ferrous) core of an LVDI or LVDT (see Section 3.2.2), by virtue of fitting the coil assembly around the coupling (with suitable clearance), then the signal generated will be a measure of the transmitted torque.

The torque transmitted by a shaft may also be measured by loading the shaft with a *dynamometer*, which is an electric generator or a hydraulic pump which must be suitably loaded in a controllable manner. If the frame of the generator (or case of the pump) is mounted in journal bearings and restrained by springs of known spring rate, then the extension (or compression) of the springs will be a measure of the torque transmitted, and can be instrumented with a displacement transducer. Such an arrangement is often used in test beds for motors, engines, and turbines.

4.4.2   Tension transducers

Tension in wires, yarns, cable, tape, sheets, filaments, felts, paper webs, plastic webs, etc. may be instrumented by using the material to support a light roller. The roller is restrained by a spring and the motion is damped by means of a dashpot (which generates an opposing force proportional to velocity), as shown in Fig. 4.19. The extension of the spring can be shown to be proportional to the tension being measured, in the steady state. This extension may be instrumented with a rectilinear displacement transducer.

Alternatively, if the material is being pulled by a driven pulley, then the tension may be inferred by measuring the torque in the drive shaft from the motor to the pulley.

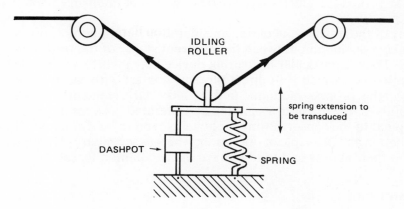

Fig 4.19   Instrumenting an idling roller in order to measure tension

5

Transducers –
Temperature, Pressure, Flow, Level, Density, pH, Humidity, Moisture and Thickness

5.1 Introduction

In this chapter, we are concerned with transducers for the measurement of variables which occur typically in production processes. Thus, the process may be creating some continuous product, and the variable to be measured might well be a major influence upon, or a direct measure of, the quality of that product. A range of transducers known loosely as *on-stream analysers*, and not covered in this chapter, tend to be more sophisticated in their nature and involve the application of complex mesurements to the analysis of the process product: these are dealt with separately in Chapter 6.

With many of the transducers under consideration here, one very often needs to make a distinction between the different stages of the transduction process. Thus, a particular measuring device may involve a 'primary element' or 'sensor', which is in direct physical contact with, or in close proximity to, the process medium, and a 'secondary element', which causes a useful display or control signal to be generated, as a result of the stimulus applied to the sensor. Primary elements tend to be distinctive in their construction and principle of operation, whereas secondary elements are limited in their variety and tend to have many common features.

5.2 Transmitters

A *transmitter* is a device for converting the output from a primary element into a useable signal, which is then transmitted either to an indicating instrument or to a controller. Thus, transmitters can be regarded as being forms of secondary element. The most common output signals generated by transmitters are either pneumatic pressure signals, in the 3 to 15 psi (20 to 100 kPa) gauge pressure range, or DC currents, in the 4 to 20 mA range. However, this latter, electrical, range is not exclusive, and alternative DC current, DC voltage, AC voltage, and digital transmissions are sometimes used.

The properties associated with using a current as an electrical signal for transmission purposes were defined in Section 2.9, and are summarised as follows:

- very often, the power source for the transmission circuit can be sited within the receiver, so that only a two-wire connection then need be used

- several loads at varying locations can be connected in series, up to a specified limit on the total resistance, with the transmitter behaving as a current source

- the length and resistance of the transmission circuit does not affect the signal sensitivity, provided that the upper limit on circuit resistance is not exceeded.

The use of an offset datum (typically 4 mA) enables a high signal/noise ratio to be preserved at all times and also enables the complete absence of a signal to be interpreted as an indication of equipment failure.

Where the signal from the primary element is a (process or instrument) pressure, the most common types of transmitter used as the secondary element are either pneumomechanical or electromechanical. The primary element may convert a single pressure into a force (for example, when the element is a Bourdon tube) or it may develop a differential pressure (for example, when the element is an orifice plate): in the latter case, the transmitter often would be referred to as a *D–P Cell* or a *ΔP Cell*.

Mechanical transmitters may be of the position balance or force balance type. The former is operated by the displacement of the primary element from a null position; the latter operates on the 'force balance' principle, described in Section 4.4.1, whereby an unknown force, the value of which is related (by the primary element) to that of the measured variable, is applied to a beam arm and is counterbalanced by a force of known magnitude.

A pneumatic force balance pressure transmitter is illustrated in Fig. 5.1. The force beam will be in equilibrium when the force developed by the output air pressure, acting in the feedback bellows, cancels the force developed by the primary element (in this case, the Bourdon tube), together with the datum-offsetting spring forces. An electronic force balance differential pressure transmitter is shown in Fig. 5.2. The force beam will be in equilibrium when the force developed by the output current I_o, causing the feedback coil to be attracted to the magnet, cancels the force due to the differential pressure $\Delta p = p_2 - p_1$ acting upon the diaphragm capsule, together with the force due to the datum-offsetting spring.

With both transmitters, the instrument behaves as a closed loop control system. In the steady state, the output signal magnitude should follow a linear relationship with the transmitter input signal generated by the primary element. With the pneumatic transmitter, the flapper and nozzle form a displacement error sensor and error amplifier (see Section 9.2.2) whilst, with the electronic transmitter, the armature and electromagnetic detector act as a component in a displacement error-sensing circuit.

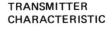

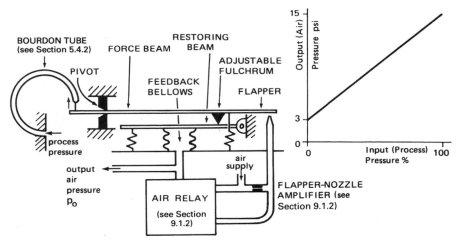

Fig 5.1   Schematic diagram of a pneumatic force balance pressure transmitter

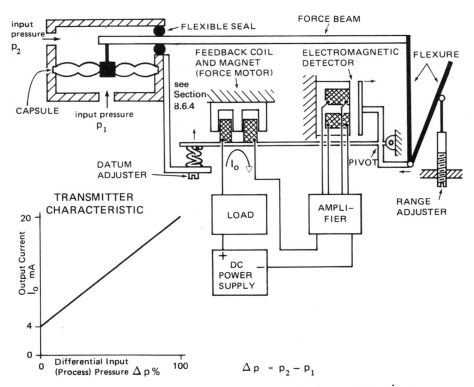

Fig 5.2   Schematic diagram of an electronic force balance differential pressure transmitter

Some non-mechanical transmitters are also manufactured and many of these involve integrated circuit technology. The advantages arising from the use of solid state technology include:

- fast response
- high linearity
- low power requirement
- physical compactness.

In some of these transmitters, the circuit may include a Wheatstone bridge, with the primary element supplying one or more limbs of the bridge: this technique is used, for example, with some temperature and pressure measuring systems. In a few cases, the bridge is constructed from strain gauge elements diffused into a single silicon crystal chip.

5.3   Temperature transducers

There are many methods by which temperature may be measured. Thermometers may be classified as follows.

Expansion thermometers

- solid expansion
- liquid expansion
- gas expansion

Change-of-state thermometers

Electrical transduction thermometers

- variable resistance
- semiconductor
- thermoelectric
- radiant energy

The types most suited to indication and control applications are those most capable of remote electrical transmission without appreciable transmission loss, which would create inaccuracy. As a consequence, electrical transduction thermometers are usually used wherever high accuracy is required. Primary temperature sensors (in this case, pressure thermometers) which convert temperature changes into pressure changes, which are then detected by a pressure transducer, are popular for some applications because they are reliable and relatively inexpensive, if less accurate.

Various solid expansion thermometers are available, the most common being the bimetal strip type. This device utilises the different coefficients of expansion of different metals which have been bonded together, thereby causing deformation when the strips are heated. Bimetal strips may be in straight, spiral or helical configurations. Other solid expansion types are the solid rod thermostat and the hot wire vacuum switch.

The most common liquid expansion thermometer is the mercury-in-glass type. For higher temperature measurement, mercury-in-steel thermometers may be used. These devices are connected to a Bourdon tube (see Section 5.4.2), which measures the pressure developed by the mercury as a result of thermal expansion. Several alternative filling liquids may be used, depending upon the temperature range and cost considerations. These thermometers are sometimes referred to as *filled thermal systems*, and will be described in greater detail in Section 5.3.4.

Gas expansion thermometers utilise the expansion of gases which results from increasing temperature. There are two types: the 'constant pressure' type, which develops an increasing volume, and the 'constant volume' type, which develops an increasing pressure, as the sensed temperature increases.

Change-of-state thermometers are basically liquid expansion thermometers in which the liquid is allowed to vaporise partially in the thermometer. The temperature is then measured in terms of the vapour pressure developed. These thermometers are often used because they are cheaper than mercury-in-steel thermometers and because they are not affected so adversely by changes in ambient temperature away from the thermometer bulb. Their useful temperature ranges are considerably narrower than those for mercury-in-steel types, however.

In electrical transduction thermometers, temperature is converted into an electrical quantity: that is, resistance, current, or voltage. The most commonly used examples of this class of thermometer are the *thermocouple*, *resistance thermometer*, *thermistor*, and semiconductor temperature sensors, and these will be described in detail later in this section.

5.3.1 Bimetallic temperature sensors

Bimetallic strips are made from two dissimilar metals, welded together, having different values for the coefficient of expansion with temperature. The two commonly used metals are brass, with a relatively high coefficient of expansion, and Invar (an alloy of nickel and iron), with a relatively low coefficient of expansion.

Usually, the strip is straight at 'normal' temperature and becomes increasingly bent as its temperature is raised, with the brass side becoming convex. In most cases, the strip is arranged to either make or break an electrical contact when a preset temperature is reached. With some designs, the strip is prestressed (by now being bent at normal temperature), with the result that a snap action occurs when the spring force to which the strip has now been prestressed is overcome by the force due to thermal expansion. Adjustment of the operating temperature is achieved by physically adjusting the contact clearance or, where relevant, the

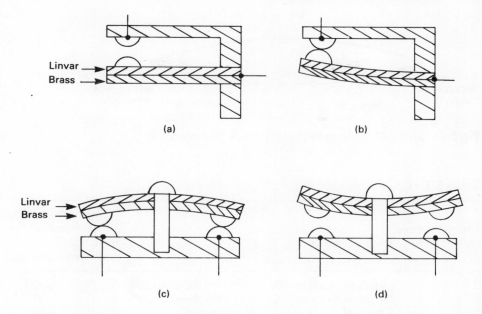

Fig 5.3   Some bimetallic strip configurations: (a) and (b) show a nominally straight strip, with (a) representing a cool state and (b) a hot state; (c) and (d) show a prestressed strip with (c) representing a cool state and (d) a hot state

degree of prestressing of the strip. Figure 5.3 shows the principles involved. Sometimes, bimetallic strips are bent into a spiral form, with one end anchored and the other connected either to an instrument pointer or to the moving member of a displacement transducer. In this manner, they behave similarly to the filled system Bourdon tube and yield a continuous indication of temperature.

5.3.2   Thermocouples

The thermocouple is classified as a thermoelectric temperature transducer, and Fig. 5.4 demonstrates the construction of a simple version. The principle of operation is based upon the Seebeck effect, which states that, whenever two dissimilar metals are connected together as shown and the junctions are subjected to different temperatures, an EMF e is generated. Moreover, this EMF is approximately proportional to the difference between the temperatures of the two junctions.

The wires of a thermocouple may be connected together to form a junction, by twisting, clamping, soldering, or (electric or gas) welding, with the last two methods being the preferred alternatives for good industrial practice. The choice of wire materials depends upon the temperature range to be sensed by the hot junction, amongst other considerations, and tends to be limited to the combinations listed in Table 5.1. Other

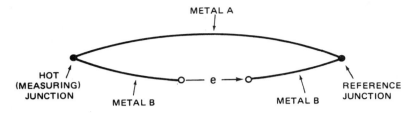

Fig 5.4   Symbolic representation of a simple thermocouple

Table 5.1   Commonly used thermocouple materials

Thermocouple type	Wire materials	Typical useful temperature range °C minimum	maximum
B	platinum rhodium 6% – platinum rhodium 30%	38	1800
C	tungsten rhenium 5% – tungsten rhenium 26%	0	2300
E	Chromel – Constantan	0	982
J	iron – Constantan	–184	760
K	Chromel – Alumel	–184	1260
R	platinum – platinum rhodium 13%	0	1593
S	platinum – platinum rhodium 10%	0	1538
T	copper – Constantan	–184	400
–	platinum 30% rhodium– platinum 6% rhodium	0	1780
–	iridium 40% rhodium – iridium	0	2000
–	tungsten – rhenium	0	2220
–	tungsten – tungsten 26% rhenium	0	2330

Note   Alumel   – nickel-aluminium alloy
Constantan   – copper-nickel alloy
Chromel   – nickel-chromium alloy

selection factors include physical strength, corrosion resistance, electrical resistance and cost.

It will be seen that the useful temperature range is very dependent upon the types of wire material used, with the lower range corresponding to base metals and base metal alloys. The wires are manufactured in a range of gauges, and are usually supplied as sleeved pairs, with the sleeving frequently colour coded for ready identification.

Figure 5.5 shows some typical calibration curves for the thermocouple types listed in Table 5.1. It will be seen that some of the curves depart significantly from straight line relationships, with the departure being most prominent in certain temperature ranges. EMF versus temperature tables are published by various standards institutions, and the data therein, together with the published calibration curves, assume that calibration has

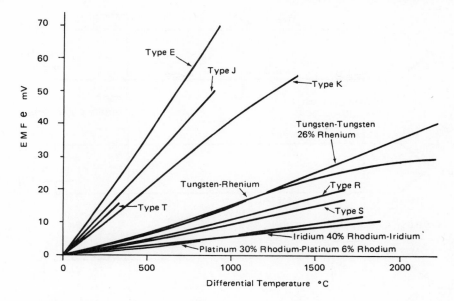

Fig 5.5   Calibration curves for commonly used thermocouple materials

occurred with the reference junction held at 0°C. Typical tolerance figures on published data range from ±1/4% to ±2% of measured temperature.

Some temperature controllers include provision for analog or digital computation of correction laws to linearise the voltage vs temperature relationship, especially for the most common types of base metal thermocouple.

Because of the small magnitude of the EMF generated, care must be taken to preserve a good signal/noise ratio. Wherever feasible, it is preferable that the voltage should be suitably amplified at a point as close as possible to the reference junction, before the signal is transmitted to a temperature controller, indicator, or recorder. It is also preferable that the reference junction should be as close as practicable to the measuring junction.

Because the thermoelectric EMF is a function of the temperature differential between the two junctions, the temperature of the measuring junction can only be interpreted from the EMF if the temperature of the reference junction is known and preferably is held constant.

Because of the various technical and cost factors, it is often impracticable to use extensive lengths of thermocouple wire. For this reason, specified types of extension wire (with alternative insulation materials), together with copper interconnections, may be interposed between the thermocouple and its load. Figure 5.6 shows some possible alternative connection configurations.

Typically, the composite reference 'junction' would be in thermal contact with a metal heat sink, but electrically isolated from it. Often, this heat sink would be shared with a set of similar thermocouple circuits. The

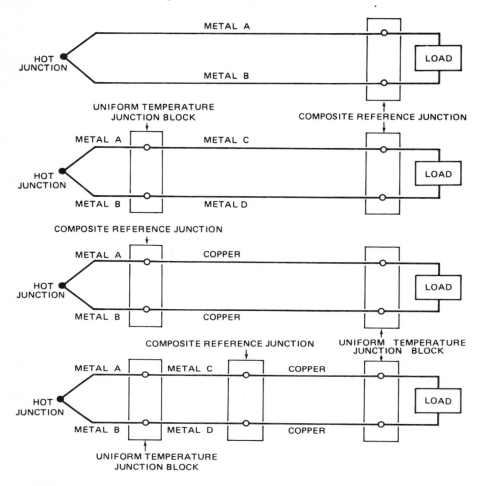

NOTE: Metals A and C are chosen to have comparable thermoelectric properties over a specified
temperature range.
Metals B and D are chosen to have comparable thermoelectric properties over a specified
temperature range.

Fig 5.6   Alternative techniques for connecting between a thermocouple
reference (hot) junction and the load component

heat sink would have its temperature measured by, for example, a
resistance thermometer (see Section 5.3.3) and the signal developed
would either be used in a temperature feedback loop to control the
temperature of the heat sink, using a small integral electrical heater, or
would be used for compensation purposes to correct the thermocouple-
generated data for sensed changes in reference junction temperature.

In order to determine the relationship between the hot junction EMF
and the voltage developed across the load, in such circuits, see Fig. 5.7.

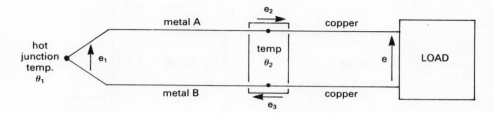

Fig 5.7 Thermocouple and load with copper conductors

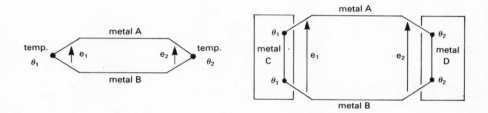

Fig 5.8 The connection of additional metals into a thermocouple circuit

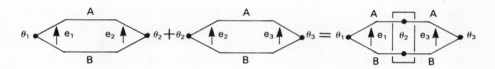

Fig 5.9 The creation of additional junctions in a thermocouple

Metals A and B have junction temperatures θ_1 and θ_2 as shown. The EMF e is not altered if either or both junctions are opened and one or more metals are interposed between metals A and B, if θ_1 and θ_2 both remain constant. This is called the Law of Intermediate Metals, and is demonstrated in Fig. 5.8. This law indicates that measuring equipment connected to the cold junction (at temperature θ_2) has no effect, provided the two points of connection are at the same temperature. Similarly, equipment connected into the hot junction (at temperature θ_1), for calibration purposes, should also have no effect.

The Law of Intermediate Temperatures states that the EMF of a thermocouple with junction temperatures θ_1 and θ_3 is the sum of the EMFs of two thermocouples, made from the same metals as the original one, and having junction temperatures of θ_1 and θ_2 for one and θ_2 and θ_3 for the second. This is illustrated in Fig. 5.9. This law is useful for calculating the effect upon the EMF of a change in the temperature of the cold (reference) junction.

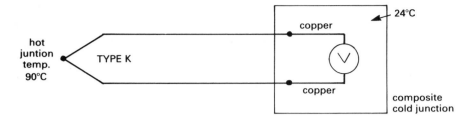

Fig 5.10   Thermocouple circuit with no cold junction compensation

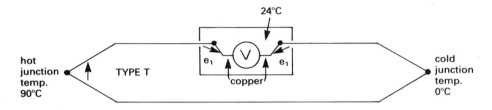

Fig 5.11   Thermocouple circuit with cold junction compensation

Figure 5.10 shows the use of a voltmeter to measure a thermocouple EMF, with no cold junction compensation. From tables, $e_{90°C} = 3.69\,\text{mV}$ and $e_{24°C} = 0.96\,\text{mV}$. The insertion of the voltmeter will follow the Law of Intermediate Metals, so that the measured voltage will be $e_{90°C} - e_{24°C} = 2.72\,\text{mV}$.

Figure 5.11 shows the use of a voltmeter to measure a thermocouple EMF, with cold junction compensation. From tables, $e_{90°C} = 3.78\,\text{mV}$, so that the insertion of the voltmeter will follow the Law of Intermediate Temperatures. Thus, the measured voltage will be $3.78 - 0 + e_1 - e_1 = 3.78\,\text{mV}$.

Because its thermal mass usually is very small, the hot junction can have a very fast response, typically in the millisecond range, to changes in temperature. However, for physical and chemical protection, the hot junction often has to be enclosed in a rigid tube (sometimes called a 'thermowell'), which is usually made either from metal (typically, stainless steel) or ceramic. This tube will add considerably to the thermal mass of this composite temperature sensor, so that the overall response time is likely to be in the range of many seconds.

Unprotected, the hot junction can be used to sense spot temperatures. When mounted within a tube, it will tend to sense the average temperature of the extremity of the tube.

Figure 5.12 shows a 'thermopile', which is a series-connected set of thermocouple junctions and which develops a voltage equal to the product of the EMF generated by one junction pair multiplied by the number of such pairs. Typically, this would be used as the sensor in a *radiation pyrometer*: this instrument is usually used to measure the surface temperature of a physically inaccessible heat-radiating body (for example, a molten ingot inside a furnace). A proportion of the thermal radiation is arranged to be transmitted through a system of windows and lenses, and focussed onto the sensor. The temperature of the sensor will settle to a steady value, which will be a measure of the temperature of the radiating body.

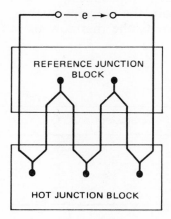

Fig 5.12   Construction of a thermopile

The advantages of thermocouples are:

● high temperature range, if appropriate metals are chosen
● small size, if unprotected
● fast response, if unprotected
● relatively linear, especially over specific temperature ranges
● low cost, especially if unprotected.

The disadvantages of thermocouples are:

● the need for compensation for changes in cold junction temperature or for control of this temperature
● low output signal level
● limited accuracy, even with precision cold junction temperature compensation or control
● the need for special extension wires, in some circumstances
● the difficulty experienced with distinguishing between an open circuit fault, in the thermocouple, and zero temperature differential.

5.3.3   Resistance thermometers and thermistors

A *resistance thermometer* is an element wound from metallic wire, and it senses temperature by virtue of the dependence of the metal resistivity upon the value of the temperature. Sometimes, it is known as a *resistance temperature detector*, or *RTD*. The most commonly used wire materials are listed in Table 5.2.

RTDs are manufactured in nominal values ranging from tens of ohms to several kilohms, with hundreds of ohms being commonplace. Typically, the element is encapsulated in a small solid glass rod, producing what is sometimes called a 'resistance bulb'; in addition, the element may be installed within a metal or ceramic tube, similar to the thermowells in which thermocouples are frequently enclosed.

Table 5.2   Commonly used RTD materials

Material	Temperature range, °C	Temperature coefficient at 25°C, /°C
nickel	− 80 to +320	0.0067
copper	−200 to +260	0.0038
nickel-iron	−200 to +260	0.0046
platinum	−200 to +850	0.0039

Figure 5.13 shows calibration curves for three of the resistance materials: it will be seen that platinum yields by far the most linear characteristic, even over a wide temperature range, and this factor, together with the advantage of the wide useful temperature range, accounts for the popularity of this material, despite its high cost. In general, the element resistance R_θ will be related to its temperature θ by a law of the form

$$R_\theta = R_o[1 + \alpha\theta + \beta\theta^2 + \gamma\theta^3].$$

α, β, and γ are constants, and the values of β and γ may be very small. R_o is the resistance at 0°C.

Thermistors consist of small pieces of ceramic material made by sintering mixtures of oxides of chromium, cobalt, copper, iron, manganese, nickel, etc. They are moulded, in various sizes, into a number of different shapes, such as beads, discs, rods, etc. and are finally encapsulated, with copper leads, typically with a colour coded vitreous material. The nominal resistance of a thermistor can vary from tens of ohms to several megohms, depending upon the type of material and the physical dimensions. The resistance versus temperature relationship is highly nonlinear, being defined by a power law of the form $R_T = ae^{b/T}$, where a and b are constants over a small range of temperature and T is absolute temperature. The majority of thermistors are of the negative temperature coefficient (NTC) type, for which R_T falls (nonlinearly) with increasing

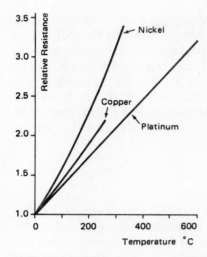

Fig 5.13   Calibration curves for three RTD materials

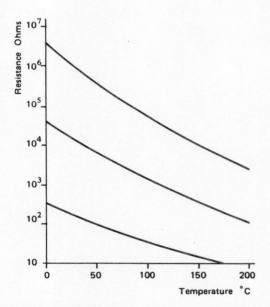

Fig 5.14   Calibration curves for three different thermistors with negative temperature coefficients

temperature; however, a few positive temperature coefficient (PTC) types also are made. Figure 5.14 shows typical resistance vs temperature curves for three different examples: although highly nonlinear, they can be much more sensitive to temperature change than comparable RTDs, within a specified temperature range. In addition they can be much cheaper.

RTDs and thermistors are normally connected as one element in a Wheatstone bridge, a typical arrangement being shown in Fig. 5.15. The 'three-wire' method of connection between the temperature sensor and the bridge is used in order that the change in lead resistance with change in ambient temperature does not affect unduly the law of the bridge: this requires that the two outer leads be identical electrically. A four-wire method of connection is also used, as an alternative. For the arrangement of Fig. 5.15, V_o is related to the applied voltage V_A and the sensor resistance R_θ by the law

$$V_o = \frac{V_A}{2} - V_A \cdot \left[\frac{R_1 + r}{R_1 + 2r + R_\theta} \right]$$

where r is the lead resistance. Thus, V_o is a (nonlinear) measure of the difference between R_θ and R_1 so that, in a temperature controller, R_1 could be set manually to equal that value of R_θ corresponding to the desired value of temperature: at the null, when $V_o = 0$, R_θ would be equal to R_1.

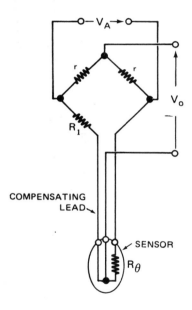

Fig 5.15 Typical circuit for use with RTDs and thermistors

The applied voltage V_A must be given a reasonably low value, so that the 'self-heating' effect within the temperature sensor is minimised. This effect can result from internally generated heat, due to $I^2 R$ dissipation, so that the temperature rise detected by the sensor resistance change would then be due only partly to the heat transmitted from the surrounding medium.

5.3.4   The filled-system thermometer

The Bourdon tube is a device for converting a pressure within the tube into a displacement at the end of the tube, and is described in Section 5.4.2. When used in a temperature transducer, the pressure is developed, by the temperature of the thermometer bulb, within the liquid, gas, vapour, or mercury which has been used to fill the system: that is, the bulb, connecting capillary, and Bourdon, as shown in Fig. 5.16. A change in pressure

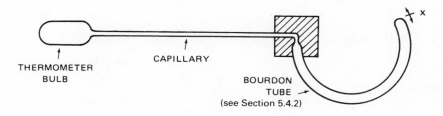

Fig 5.16   Construction of a filled-system thermometer

causes a physical deformation of the Bourdon, resulting in a displacement x at its free end: this displacement can be made to be proportional approximately to the temperature change at the bulb. The device is robust and cheap but not particularly accurate, especially if means are not included for compensating for the effect of changes in ambient temperature upon that part of the fluid circuit within the capillary and the Bourdon. Filling media include alcohol, xylene, mercury, nitrogen, and hydrogen, selectively covering a temperature range from $-120°C$ to $+650°C$.

5.3.5   Semiconductor temperature transducers

Over a specified temperature range (typically, $-55°C$ to $+150°C$), silicon junction semiconductor devices are well suited to temperature measurement, being fast, highly linear, and cheap. The junction potential of silicon diodes and transistors, although varying between devices, changes at about 2.2 mV/°C.

A typical two-terminal sensor is made using a junction transistor, as shown in Fig. 5.17 with a representative calibration graph. In order that variation in device current should not be allowed to affect the calibration, this type of device should be supplied from a constant-current DC source and the voltage dropped across it should be sensed by a buffer amplifier, as shown in Fig. 5.18.

Other, more sophisticated, semiconductor temperature sensors have been designed to function as current sources: a typical IC device develops a current in μA numerically equal to the temperature, in Kelvin, of the case. A typical application is shown in Fig. 5.19. Yet another type behaves

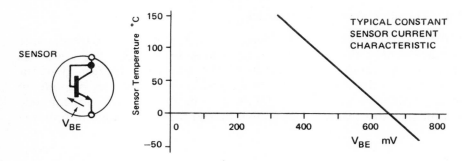

Fig 5.17   Calibration graph for a junction transistor used to sense temperature

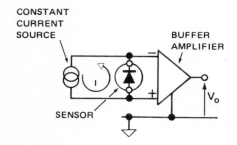

Fig 5.18   Typical circuit employing a voltage-generating temperature-sensing semiconductor

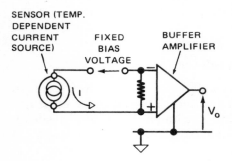

Fig 5.19   Typical circuit employing a current-generating temperature-sensing IC

as a temperature-dependent zener diode, with the zener voltage related to the temperature with a sensitivity of 10 mV/K. Some commercially available devices incorporate the temperature sensor, buffer amplifier, and possibly the DC power source, into a single IC package, having a typical accuracy of 1°C.

5.4   Pressure transducers

Pressure is defined as force per unit area and the fundamental unit in the SI system is the Newton per square metre (N/m^2), also known as the Pascal (Pa). Because the N/m^2 is an inconveniently small unit, the kN/m^2 (kPa) and the Bar ($100\,kN/m^2$) are more commonly used. Low pressures are often expressed in millimetres of water (mm H_2O) or millimetres of mercury (mm Hg). It is important to note that the liquid column is not a true pressure unit, since it does not have the dimensions of force/area. A liquid column is related to pressure by the expression $p = \gamma h$, where p is the pressure in N/m^2, h is the height of the column in metres, and γ is the specific weight of the liquid in N/m^3.

Pressure measured from the true zero pressure is termed 'absolute', and that measured relative to the local atmosphere is termed 'gauge', so that

absolute pressure = gauge pressure + atmospheric (barometric) pressure

Pressures measured relative to atmospheric pressure may be either positive or negative. Where the latter is the case, the word 'vacuum' is often used to indicate an amount of pressure below the atmospheric value: for example,

$$10\,kN/m^2 \text{ vacuum} = \text{atmospheric pressure} - 10\,kN/m^2$$

on an absolute scale; that is, approximately $100 - 10 = 90\,kN/m^2$ absolute. In vacuum measurements, the Torr unit is often used, where 1 torr = 1 mm Hg.

Most pressure transducers generating an electrical output incorporate one of the following elements as the primary pressure sensor, which converts the pressure into a displacement or a force:

- manometer
- Bourdon tube
- bellows
- diaphragm.

Most of the devices used to convert the resulting displacement or force into an electrical signal have been discussed in Chapters 3 and 4, and may be either strain gauge, capacitive, inductive, variable reluctance, piezoelectric, or variable differential transformer types.

A recently introduced type of pressure transducer, utilising a piezoelectric sensor in conjunction with an integrated circuit, is illustrated in Fig. 5.20. The silicon sensor, which responds to either positive or negative pressure differentials, includes a reference pressure cavity which encloses a vacuum, in absolute transducers, or is open (via the alternative input port), in gauge and differential transducers. With minor circuit modifications for the differential and backward gauge versions and inclusion of appropriate input ports, the same basic transducing device can serve all three types of pressure measurement. The accuracy of this type of trans-

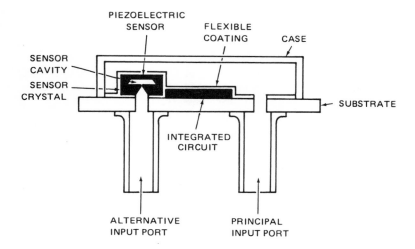

Fig 5.20 Pressure transducer using a piezoelectric sensor and integrated circuit

ducer is limited to about 1½% and it is unsuitable for measurement of very low pressure differentials, but it has an extremely fast response to pressure changes.

5.4.1 Manometers

Liquid manometers are simple pressure sensors which are economical, reliable, and accurate. There are two types: the visual (sometimes called a 'sight glass' or 'sight gauge') and the float. The latter type is necessary for high-pressure service and/or where transduction is desired, and is illustrated in Fig. 5.21: the float displacement will be converted into an electrical signal by one of the displacement transducers of Section 3.2. The use of this type of instrument is declining.

5.4.2 Bourdon tubes

Bourdon tubes are elastic deformation elements and, because of their simple design and low cost, they are used more widely than any other type of pressure sensor. There are three types of Bourdon element: the 'C', the spiral, and the helical. These are illustrated in Figs 5.1, 5.22 and 5.23 respectively. Bourdon tubes are made from metals such as phosphor bronze, beryllium copper, stainless steel, Monel, and certain steel alloys. Pressure ranges up to 100 000 psi (680 MPa) are quoted. Typical accuracy is in the range of ±½% to ±1% of span.

   The advantages of Bourdon tubes are:

● low cost
● simple construction

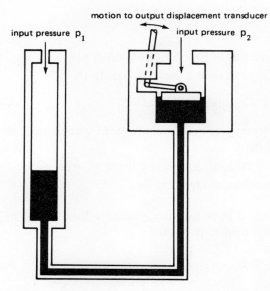

Fig 5.21   Float type of liquid manometer

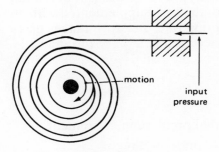

Fig 5.22   Spiral Bourdon tube

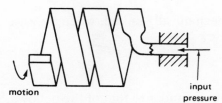

Fig 5.23   Helical Bourdon tube

- high reliability
- high pressure ranges
- good accuracy versus cost, except at low pressure ranges
- suitability for transducers for generating electrical outputs.

The disadvantages of Bourdon tubes are:

- low spring rate and therefore limited measurement precision, at pressures below about 350 kPa
- susceptibility to shock and vibration, due to the large overhang
- poor repeatability, due to mechanical hysteresis.

Bourdon tubes need to be operated in pairs, mechanically linked, when they are required to measure differential pressure.

5.4.3   Bellows

A bellows is an elastic deformation element usually formed from a thin seamless tube. A simplified representation of the element, used as an indicator, is shown in Fig. 5.24. The greatest use for bellow units is as receiving elements for pneumatic recorders, indicators, and controllers. Bellows are made from metals such as brass, phosphor bronze, beryllium copper, stainless steel, and Monel. Often, they are supplemented with an integral helical spring, to modify their spring rate. Pressure ranges up to 400 psi (2.7 MPa) are common. Typical accuracy is $\pm\frac{1}{2}\%$ of full span.
   The advantages of bellows are:

- high delivered force
- moderate cost
- adaptability to absolute and differential pressure measurement
- good accuracy in the low-to-moderate pressure range.

The disadvantages of bellows are:

- ambient temperature compensation requirement
- unsuitability for high pressures
- limited range of materials.

Bellows need to be operated in pairs, mechanically linked, when they are required to measure differential pressure.

5.4.4   Diaphragms

The operating principle of the diaphragm is similar to that of the bellows. Pressure is applied to the element, causing it to deform elastically, in direct

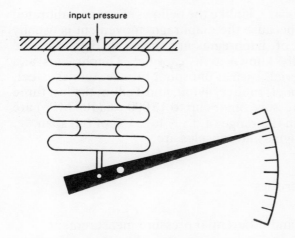

Fig 5.24 Bellows used for pressure indication

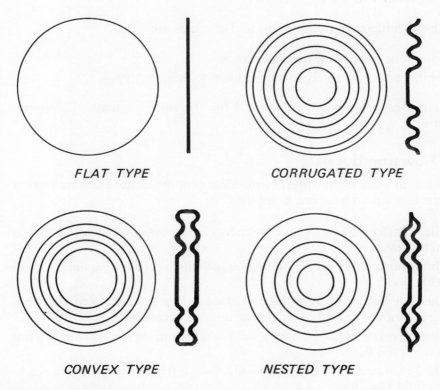

Fig 5.25 Four different examples of pressure-sensing diaphragms

proportion to the applied pressure. Unlike the bellows element, calibrated springs are rarely required, because the diaphragm movement is usually small. Some typical designs of diaphragms are shown in Fig. 5.25: the double-walled type is usually known as a 'capsule'. Diaphragms and capsules are made from materials such as phosphor bronze stainless steel, beryllium copper, nickel, Monel, rubber, nylon, and Teflon: the last three apply only to diaphragms. Pressure ranges up to 15 000 psi (100 MPa) are quoted. Typical accuracy is in the range of $\pm\frac{1}{2}\%$ to $\pm1\%$ of full span.

The advantages of diaphragms and capsules are:

- moderate cost
- good overrange characteristics
- good linearity
- adaptability to absolute and differential pressure measurement
- availability in materials with good corrosion resistance
- smallness in size
- adaptability to use with slurries.

The disadvantages of diaphragms and capsules are:

- poor shock and vibration resistance
- limitation to relatively low pressures, with many types.

Evacuated capsules need to be used for the measurement of absolute pressure.

5.5   Flow transducers

There are at least eight different physical properties used for measuring the flow of fluids. These are as follows.

- The transformation of kinetic energy into pressure energy, which is then measured.
- The generation of an electrical voltage which is proportional to linear velocity.
- The transformation of the linear velocity of the fluid into a corresponding rotational velocity, which can then be measured.
- The inference of fluid velocity from its effect upon the cooling of a hot body in the fluid.
- The generation of fluid oscillations, with a frequency proportional to the fluid velocity.
- The inference of fluid velocity from its effect upon the velocity of sound in the moving fluid.
- The use of some form of tracer, to detect linear velocity of the fluid.

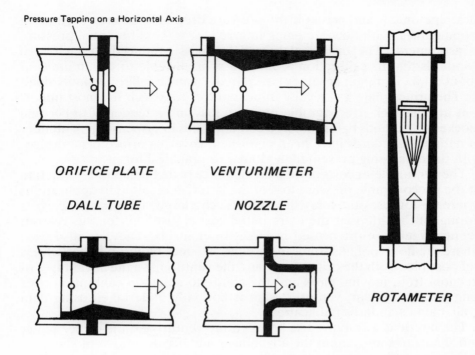

Pressure Tapping on a Horizontal Axis

ORIFICE PLATE   *VENTURIMETER*

DALL TUBE   *NOZZLE*

ROTAMETER

Fig 5.26   Five of the most common types of flow rate sensor

- The direct measurement of the total quantity flowing, using a positive displacement device.

Each principle can be applied to a number of different methods.

Fluid flow can be measured either as flow rate or flow volume. Flow rate is the integrated velocity of the individual streamlines, which make up the total velocity profile across the pipe. Flow rate measurement devices can be used to provide a direct visual indication or the transmitted output can be fed to remote indicators, recorders, or automatic controllers. Flow volume devices measure the total volume of fluid which has passed through a pipe in a given interval of time: typically, they are used for fiscal monitoring purposes, but rarely as feedback transducers. The most common types of flow rate measurement device are shown in Fig. 5.26.

5.5.1   Orifice, venturi, Dall tube, and nozzle

The most commonly used types of flowmeter for process control applications are the orifice plate and the venturi (meter). The orifice plate owes its popularity to its simplicity and the fact that it is inexpensive. It can be manufactured cheaply from metal plate and inserted between the flanges of a pipe union, and across it the pressure drop may be measured using tapping lines, with flowrate being inferred from the differential pressure measurement. The orifice plate can be mass produced to such high standards that its calibration coefficient is predictable to within limits of about ±1%; if greater accuracy is required, it can be calibrated individually. The orifice plate has several disadvantages, of which the most important are the high irrecoverable pressure loss and a tendency to become blocked if the fluids being metered contain solids in suspension. A number of national standards have been published, covering orifice plate design, with the data being presented in tabular or graphical forms.

The venturimeter costs more than an orifice plate but has only a fraction of the irrecoverable pressure loss of the latter. It is suited to applications where the low pressure loss will result in such a large saving in energy costs throughout its life that the extra initial cost is justified; for this reason, venturimeters are often used in large water mains. They are also used where the flow contains solids in suspension, since build-up of particles is not possible. Both the orifice plate and the venturi have the advantages of freedom from moving parts and the ability to maintain calibration over long periods. Again, venturi design is normally undertaken using data available in standards publications.

The nozzle is a compromise between the venturi and the orifice plate, and shares to some extent the advantages and disadvantages of each.

Devices such as the Dall tube combine the lower pressure loss and smoother flow path advantages of the venturi with a shorter, simpler construction. The Dall tube resembles a miniature venturi and it is inserted within the pipeline.

These four types of flowmeter, which all feature the insertion of an obstruction in the pipeline, suffer from the disadvantage that the measured differential pressure is proportional to the square of the inferred flowrate, so that, when the flow is (say) at one third of maximum, the differential pressure is only one ninth of maximum. With the flowmeter connected to a differential pressure transmitter, the output signal can be processed by means of a 'square-root extractor', in order to restore the relationship between the generated signal and the flowrate to a linear law.

5.5.2   Rotameters

Rotameters are variable area flowmeters which must be inserted in vertical pipelines in which the flow is directed upward. The rotameter consists of a glass or metal tube which is tapered on the inside, with the narrower end at the bottom, and in this sits a heavy plummet, which is typically tapered to a point at the bottom. The flow lifts the plummet to a stable position whereby the upthrust just cancels the weight of the plummet. The displacement will be approximately proportional to the flowrate, so that the rotameter can be regarded as a linear transducer. For adaptation to signal transmission applications, the plummet may become the moving member within an LVDI or LVDT surrounding the tapered tube, if ferrous material is embedded within the plummet; alternatively, the plummet may contain a permanent magnet, with the displaced magnetic field being tracked by a moving ferrous member outside the tube. In some designs, the plummet is fluted, so that it spins with an angular velocity proportional to the flowrate, with this velocity being sensed by a suitable angular velocity transducer. The main disadvantage with the rotameter is that the calibration is dependent upon the density of the flow medium and this may not be constant; in addition, the plummet tends not to be stable for flowrates much below 10% of full range. Ranges up to 1300 gal/min (5850 l/min) are possible, with accuracy varying from $\pm\frac{1}{2}\%$ to as poor as $\pm 10\%$.

5.5.3 Magnetic flowmeters

The principle of these devices is indicated by Fig. 5.27. The liquid must have at least a minimal conductivity and acts as a (liquid) conductor moving through an electromagnetic field, so that an EMF, proportional in magnitude to the velocity of the fluid, is developed between the orthogonal electrodes. Usually, the excitation is AC, to prevent build-up of corrosive deposits on the electrodes. This flowmeter has the advantages of good linearity, of presenting zero obstruction to the liquid flow, of being bidirectional, of generating an electrical signal directly, and of having a calibration almost independent of the type of flow medium; however, it is affected by entrained bubbles.

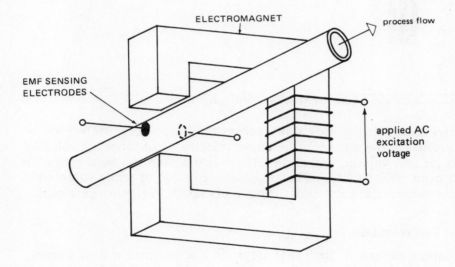

Fig 5.27 Symbolic representation to demonstrate the principle of the magnetic flowmeter

5.5.4   Turbine meters

With turbine meters, a small impeller is mounted coaxially with the centreline of a straight length of tube. The flow stream causes the impeller to spin with an angular velocity which is proportional to the flowrate, with a good linearity in the range between about 10% and 100% of rated flow, which may have a value as high as 200 000 l/min. The angular velocity of the turbine needs to be sensed, and typically this is achieved by making the impeller the rotor of a digital tachogenerator, as described in Section 3.4.5. The impeller would be made from metallic material and the pickup coils would be placed against the outside of the (non-metallic) tube. Typical accuracy is in the $\pm\frac{1}{4}\%$ to $\pm\frac{1}{2}\%$ range. The turbine meter will cause a small pressure loss to be developed, proportional to the square of the flowrate, and the impeller and its bearings may be susceptible to corrosion and entrained solids.

5.5.5   Pitot tubes

Figure 5.28 shows a cross-section through a Pitot tube, which measures linear velocity of the flow at one point in the cross-section through the flow stream. The forward-facing tap senses dynamic pressure, whilst the

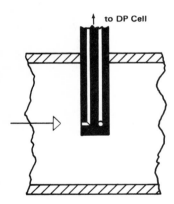

Fig 5.28   Pitot tube

sideways facing tap senses static pressure: the pressure difference is proportional to the square of the linear velocity. Pitot tubes are suitable mostly for gas, vapour, and clean liquid streams, and can be accurate to within from $\pm\frac{1}{2}\%$ to $\pm5\%$. They must be calibrated to relate volumetric flowrate to the linear velocity at the point of impact. Pressure loss is small.

5.5.6   Target meters

With target meters, a small disc target is inserted into a flow stream, mounted on a bar which passes through a flexible seal in the wall of the pipeline, as shown in Fig. 5.29. The velocity of the flow at the point of impact causes a force to be developed in the bar, and this force is proportional to the square of the velocity. The constant of proportionality will depend upon the properties of the flow medium. The force exerted upon the bar can be sensed, outside the pipeline, either by using bonded strain gauges or by nulling the force with a force-balance type of transmitter. These meters are suitable particularly for viscous, dirty, and corrosive fluids. They must be calibrated to relate volumetric flowrate to the linear velocity at the point of impact, and can be accurate within from $\pm\frac{1}{2}\%$ to $\pm3\%$.

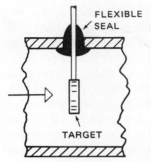

Fig 5.29 Target meter

5.5.7 Vortex meters

Vortex meters are constructed typically by inserting a fixed, fluted obstruction into a pipeline, to create a swirling flow which oscillates at a frequency proportional to the volumetric flowrate. Thermistors are mounted in this swirling flow and are energised so that the temperature developed by self-heating oscillates at the frequency of the vortex: see Section 5.3.3. The frequency of the pulsating voltage developed by the thermistor network is used as a measure of the flowrate, and can be converted into an equivalent voltage magnitude using a frequency-voltage converter. Excellent linearity can be achieved, within specified operating limits, and the accuracy can be better than $\pm1\%$.

5.6 Level transducers

Level measurement may be classified broadly into two general groups: direct and inferential. Direct level measurements are simple and economical. Usually, they are visual as, for example, sight glasses, dip sticks, and calibrated tapes, and are not adapted easily to signal generation.

Inferred methods depend upon the medium having a property which is related to level and is measurable. For this purpose, use has been made of the many physical and electrical properties which are well suited to the generation of proportional output signals for remote transmission. Included in these properties are the following.

● Hydrostatic head – the force or weight produced by the height of the liquid, which would be sensed by a pressure transducer placed at the bottom of the vessel.

- Buoyancy – the upward force of a submerged body, which is equal to the weight of the fluid which it displaces, or the upward displacement of a float on the surface. Here, strain gauges or displacement transducers would be used, in order to establish a suitable output signal.

- Conductance – at desired points of level detection, the medium to be measured conducts (or ceases to conduct) electricity between two fixed probe locations or between one probe and the vessel wall.

- Capacitance – the medium to be measured serves as a variable dielectric between two capacitor plates. Two substances form the composite dielectric: the medium whose measurement is desired and the vapour space above it. The total capacitance value changes as the volume of one material increases whilst that of the other decreases. The capacitance levels usually obtained are in the pF range. Typically, an AC bridge operated at a supply frequency in the vicinity of 1 MHz is used, in order to convert the capacitance to a voltage signal: this occurs when the change in capacitance with liquid level unbalances the bridge. The calibration of voltage vs liquid level may show a significant departure from linearity. In addition, the sensitivity will vary with dielectric permittivity and therefore with the composition of the liquid and vapour.

- Radiation – the measured medium absorbs radiated energy. As in the capacitance method, vapour space above the measured medium also has an effect upon the measurement, due to its own absorption characteristic, but the difference in absorption between the two is great enough for the measurement to be made.

- Sonar or ultrasonic – the medium to be measured reflects, or affects in some other detectable manner, high-frequency sound signals generated at appropriate locations near the test medium.

Caution must be applied, when utilising inferred level measurement, to ensure that the measured property has a well-defined relationship to level.

Normally, little difficulty is experienced with measuring the level of clean, low viscosity fluids. Slurries, viscous substances, and solids present much greater problems. For evaluation of the method to be used for a particular application, certain operating conditions, such as level range, fluid characteristics, temperature, pressure, and the state of the fluid around the operating area, must be known. Figures 5.30, 5.31 and 5.32 show examples of hydrostatic head, capacitance, and buoyancy types, respectively.

Figure 5.33 shows alternative arrangements for mounting a radiation source (S) and a radiation detector (D) for the measurement of liquid level. (Section 6.9 discusses types of radiation, radioactive sources, and radiation detectors). Because the absorption varies in inverse proportion with the thickness of the absorbing material, the relationship between detector signal and liquid level will depend heavily upon the orientation of

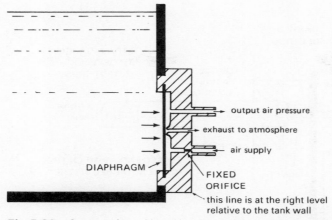

Fig 5.30 Sensing liquid level by instrumenting hydrostatic head pneumatically

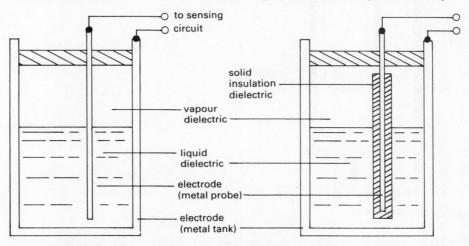

Fig 5.31 Sensing liquid level by constructing the vessel as a giant capacitor

the source and detector. These techniques may be modified to suit the measurement of levels of solids and slurries.

5.7 Density transducers

Density may be defined as mass per unit volume, and is expressed in kg/m^3 in the SI system of units. Specific gravity (which is not recommended in the SI system) is often used synonymously with density, and is defined as the ratio of the density of the fluid in question to the density of water, at a specific temperature.

Common density measurement methods involve air bubble, displacement, displacement U-tube, vibrating U-tube, and radiation techniques.

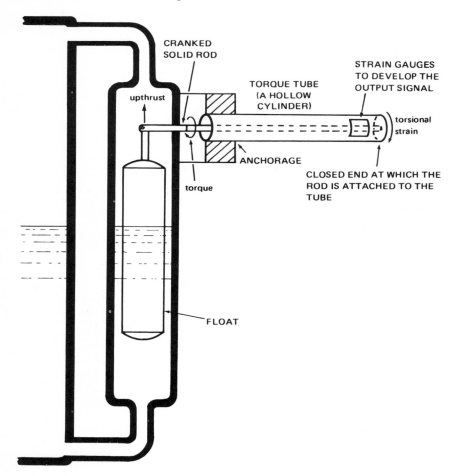

Fig 5.32   Inferring liquid level by using an instrumented torque tube to restrain the upward displacement of a partially immersed float

5.7.1   Air bubble type

The simplest and possibly the most widely used density-measuring device is two bubbler tubes set at different levels in a vessel containing liquid, as shown in Fig. 5.34. The two tubes act as back-pressure generators, and the difference between the two pressures is measured by a differential pressure transmitter and is equal to a constant-height column of the liquid. Change in differential pressure is proportional to density change.

5.7.2   Displacement type

Displacement density instruments operate using the buoyancy of a completely submerged body, as shown in Fig. 5.35. The force acting on the

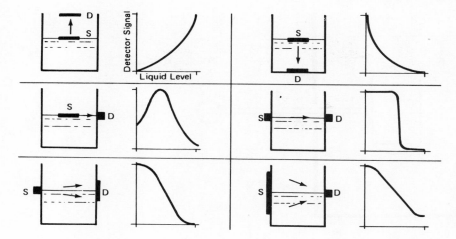

Fig 5.33 Six alternative techniques for using a radiation source and detector to sense liquid level

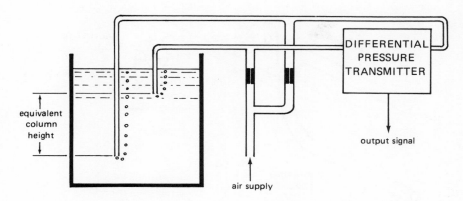

Fig 5.34 Sensing differential back-pressure to infer liquid density using bubbler tubes

balance (or torque) arm is related directly to the density of the liquid displaced by the float.

5.7.3 Displacement U-tube type

In this arrangement, the process liquid flows through a U-tube, the weight of which is balanced on a weigh beam. The weigh beam is linked mechanically to the flapper-nozzle assembly of a force balance transmitter. Change in density modifies the flapper-nozzle separation, changing the back-pressure within the nozzle and hence the transmitter output signal: see Section 9.2.2.

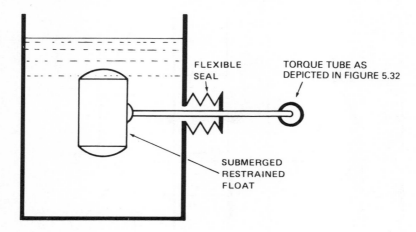

Fig 5.35   Inferring liquid density by using an instrumented torque tube to restrain the upward displacement of a fully immersed float

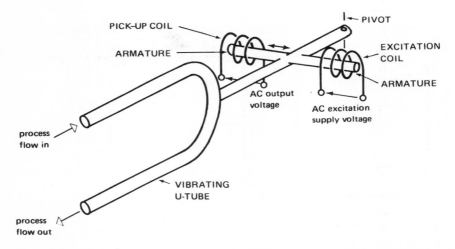

Fig 5.36   Symbolic representation of a vibrating U-tube type of liquid density transducer

5.7.4   Vibrating U-tube type

The vibrating U-type operates on the principle that the amplitude of vibration of a vibrating body is proportional to its mass. In Fig. 5.36, the total mass of the U-tube includes the mass of the flowing liquid and, therefore, changes with a change in density. An armature and coil assembly form a 'pick-up' in which an AC voltage is induced by the vibration of the amature. This voltage is proportional to the amplitude of vibration and hence to the density of the liquid.

5.7.5 Radiation type

Figure 5.37 shows the arrangement of a source, emitting either β or γ radiation, and a radiation detector, for the measurement of the density of a material flow. The absorption of the radiation by the flow medium will increase as the density of the medium increases, so that the detector will sense a corresponding reduction in received radiation. (Section 6.9 discusses types of radiation, radioactive sources, and radiation detectors.) This technique is particularly useful for density measurements on slurries.

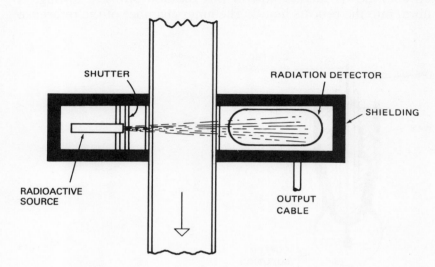

Fig 5.37 Use of a radioactive source and radiation detector to sense liquid density

5.8 pH transducers

The effective acidity or alkalinity of a liquid is normally expressed in pH. A pH of 7 corresponds to a neutral solution. The pH increases towards 14 when the alkalinity increases and approaches 0 as the acidity increases. The electrochemical reaction which takes place in a liquid when acid is added consists of an increase in hydrogen ions and a decrease in hydroxyl ions. Since the hydrogen ions are positive and the hydroxyl ions are negative, it is possible to measure the change by electrical means.

The most common method of measurement uses the glass electrode and calomel reference electrode combination. The glass electrode contains a buffer solution, which is a solution which maintains a certain pH level even when a limited amount of acid or base substance is added. The standard Beckman glass electrode, which is shown in Fig. 5.38, has a spherical end with a diameter of about 12 mm filled with the buffer solution. The wall is of very thin glass and constitutes a membrane between the buffer solution and the liquid under measurement. A platinum wire, coated with silver chloride, connects the buffer solution to the electrode head. The potential difference between the buffer solution and the process liquid, across the glass membrane, is of measurable size, despite the high membrane resistance which may be as high as 100 MΩ.

The circuit is completed through a calomel electrode, shown in Fig. 5.39, which serves as a reference electrode with a constant output EMF. A liquid junction between the process liquid and the calomel is established by means of an outer chamber, which is filled with a saturated solution of potassium chloride. A minute flow of this solution diffuses, through a porous fibre, into the process liquid. The inner chamber of the reference

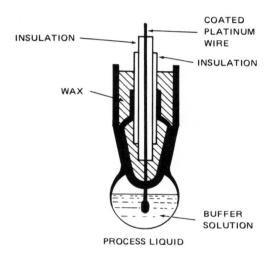

Fig 5.38 Typical construction of the Beckman electrode for sensing pH

electrode is filled with glass wool in its lower part, and on top of this is placed a layer of paste made from a mixture of calomel and mercury. A pinhole provides a liquid junction between the two chambers and completes an electrolytic cell of constant EMF.

The two electrodes are mounted in close proximity and exposed to the process liquid, although they are likely to be partly enclosed by a protective envelope. Figure 5.40 shows the connection of the electrodes to an instrumentation amplifier, which must draw a current not exceeding 1 pA, if polarisation of the cells is to be avoided.

Figure 5.41 shows calibration curves for the electrode system and indicates that the calibration is strongly temperature dependent. For this reason, some form of temperature sensor would be mounted close to the glass electrode, and the temperature data generated would be used in computation to compensate, for temperature effects, the amplifier output voltage.

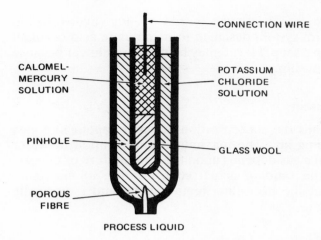

Fig 5.39 Typical construction of a reference electrode, to be used in conjunction with the Beckman electrode for sensing pH

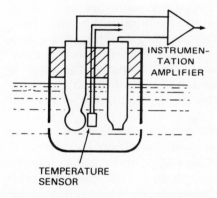

Fig 5.40 Typical assembly of electrodes and temperature sensor forming a pH transducer

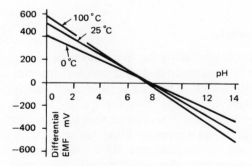

Fig 5.41 Typical calibration curves for a pH transducer for three different operating temperatures

Because the definition of pH is given by pH = −log (hydrogen ion concentration) any control system designed to manipulate acid or alkali flow, in order to maintain a set pH level, may be highly nonlinear because of this logarithmic relationship.

5.9   Humidity transducers

The term 'humidity' means the concentration of water vapour in a gas, with this gas usually being air. The maximum concentration of water vapour which can exist in a gas depends upon the temperature of the gas. 'Relative humidity' is the ratio of actual concentration to maximum possible concentration at the prevailing temperature, and is normally expressed as a percentage.

5.9.1   The hygrometer

Certain threads, such as human hair or nylon thread, are hygroscopic and therefore tend to absorb an amount, of water, which is dependent upon the prevailing relative humidity. The thread will contract or extend, depending upon whether water is released (by evaporation) or absorbed, respectively. One common type of mechanical hygrometer uses a suspended thread, and the contraction and extension of this thread is measured by a suitable displacement transducer. An alternative form uses a pneumatic force balance arrangement similar to that of Fig. 5.1, except that the source of input force now becomes the tension in the suspended thread.

Electrical hygrometers generally use two precious metal wires or grids which are precisely spaced a small distance from each other. The intervening space is filled with a layer of hygroscopic compound, such as lithium chloride, the electrical conductivity of which changes in proportion to the amount of water absorbed or released. The two wires would normally be connected to form one limb of a Wheatstone bridge, so that the output from the bridge would be a measure of the change in conductance of the path through the hygroscopic coating.

5.9.2   The wet and dry bulb thermometer

This intrument uses two thermometer bulbs, with the dry bulb measuring the ambient temperature. The wet bulb is surrounded by a porous material which is kept moist and across which a stream of the test medium is arranged to flow, using a fan if necessary. The wet bulb temperature falls in value as a result, because of evaporation, and this fall in temperature is dependent upon a combination of the ambient temperature, as sensed by the dry bulb, and the relative humidity of the gas flow. Resistance thermometers or filled system thermometers are usually used, with the output data provided by the two thermometers being employed in the computation of the relative humidity.

5.9.3 The dew point thermometer

The dew point is that temperature to which a sample atmosphere can be cooled before water condensation occurs, and is therefore a measure of the relative humidity of the sample existing before cooling is commenced.

One type of dew point instrument causes the sample gas to be drawn past a mirror, which is mounted on a refrigerated chamber. This mirror is used to reflect a light beam, and the intensity of the reflected light is reduced when condensation occurs on the mirror. The reflected light is sensed by a photosensitive detector, the output of which is amplified and used to energise a heater, which is mounted against the back of the mirror. Because condensation diminishes when the mirror is heated, the feedback action ensures that the condensation does not grow, so that the mirror temperature is held at the dew point. A temperature sensor, mounted on the mirror, is used to measure the dew point temperature.

5.10 Moisture transducers

The term 'moisture' means the concentration of liquid water in a solid material, and its measurement is particularly important in the manufacture of paper and textile webs.

One of the most commonplace moisture transducers uses a Wheatstone bridge arrangement to measure the conductance of a path through the web: this conductance is a measure of the moisture content. The web is passed through two metal pinch rollers which establish the electrical path through the web, and the resistor created by this path is connected as one limb of the bridge.

The other most commonplace moisture transducer uses an AC circuit arrangement to measure the web permittivity, which is also an indication of the moisture content. The web is passed over the faces of two separated metal electrodes, which cause the segment of web between these two faces to behave as the dielectric of a capacitor, the capacitance value of which is converted, by the circuit, into a corresponding signal level.

5.11 Thickness transducers

Thickness transducers are used for measuring the thickness of a body which, in most instances, is in motion. Where the variation in thickness is large and physical contact with the body is permissible, it may be possible to measure the position of a probe touching the surface or the separation of two pinch rollers enclosing the body, using one of the displacement transducers described in Section 3.2. In the many instances where physical contact with the body is not permitted and/or the variation in thickness is small, displacement transducers cease to be appropriate. In these cases, more sophisticated means are necessary and typical of these are techniques employing radiation sources and sensors.

Figure 5.42 shows alternative arrangements of radiation sources and

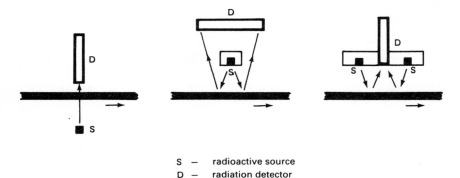

S — radioactive source
D — radiation detector

Fig 5.42 Three alternative arrangements of radiation sources and sensors to measure the thickness of a moving web

detectors for the measurement of thickness, either by direct transmission or by reflection (which is known as 'back-scattering', in this context). Radioactive sources may be chosen for the emission of α, β, γ, or X rays, depending upon the type of material being measured. α, β, γ, and X ray detectors may be used, the last named being required for 'X-ray fluorescence' detection in which X-rays at specific frequencies are reflected from the test body when it is irradiated with γ or X rays. (See Section 6.9 for a description of radiation, radioactive sources, and radiation detectors). These techniques are suitable for measuring the thickness of sheets of cardboard, paper, rubber, plastics, metal, cloth, mica, etc., of coatings of paint, of metal plating and film, and of moving beds of slurries and solid particles.

6

On-Stream Analysers

6.1 Introduction

Analysers are required for the assessment of the quality of the gaseous, liquid, and solid products of manufacturing processors. It is preferable always to measure directly the value of the property of interest but often this is impracticable, so that frequently it is necessary to infer the value of this variable by measuring related properties. Thus, for example, electrical conductivity may be measured in order to infer chemical composition. Off-line analysis in the laboratory may be undertaken to measure the property of interest but such a procedure is inappropriate for feedback control. Thus, here we require analytical instruments which function as feedback transducers, so that they must operate on-line and must generate electrical (or pneumatic) feedback signals, to be used by the relevant process controllers.

Analysers which function, automatically, on-line are referred to as *On-Stream Analysers* and these can take many forms, some of which will have off-line laboratory analyser counterparts. An on-stream analyser may be extremely expensive (say, in the order of several $10 000), when compared with other feedback transducers, and this expense means that such an analyser must be selected with great care and must be maintained meticulously. Some on-stream analysers perform their analysis on a sequence of samples of the test product, because by their nature the analytical procedure is a batch process: whenever such an analyser is used as a feedback transducer, the resulting control system will respond with the type of behaviour associated with sampled-data control systems. Another characteristic of some of these analysers, especially the sampling types, is the significant dead time which may be inherent in the analytical process: this dead time will affect adversely the stability of the associated closed loop control system.

The number of varieties of analyser falling into the on-stream category is considerable and, for this reason, only the most commonly used types can be covered here in detail. The alternative operating principles considered in this volume are as follows:

- thermal conductivity, measured in terms of the cooling effect of the test medium upon a heated filament
- combustibility, measured in terms of the heating effect of the test medium when it is ignited by a hot filament
- chromatography, which involves decomposing a mixture into its chemical component parts, by passing it through a porous material
- spectrometry, which involves decomposing a mixture into its chemical component parts, by ionising the mixture and then separating the different types of ion
- ultra-violet photometry, which involves measuring the degree of absorption of UV radiation by the test medium
- visible light photometry, which involves measuring the degree of absorption of visible light by the test medium
- infra-red photometry, which involves measuring the degree of absorption of IR radiation by the test medium
- colorimetry, which involves measuring the spectrum of visible light transmitted through, or reflected by, the test medium
- turbidimetry, which involves measuring the intensity of visible light transmitted through the test medium
- paramagnetism, which involves measuring the flowrate of those gases which are capable of being attracted by magnetic fields
- refractometry, which involves measuring the degree of refraction of light beams transmitted through the test medium
- density measurement, which involves measuring the density or specific gravity of the test medium
- pH measurement, which involves measuring the degree of acidity or alkalinity of the test medium
- humidity measurement, which involves measuring the moisture content in gases and vapours
- moisture measurement, which involves measuring the moisture content in solid materials
- radiation measurement, using radioisotope radiation sources, which involves measuring the intensity of radiation transmitted through, or reflected by, the test medium.

Table 6.1 lists these principles and relates them to the type of test medium to which they can be applied, and also indicates whether the analysis can be continuous or must be made upon samples.

6.2 Chromatographs

Chromatographs are generally used for the analysis of gases and vapours, but versions are also available for the analysis of liquids. They are amongst the most expensive on-stream analysers.

Table 6.1 Applicability of analytical principles to different test media

Analytical principle	Type of test medium – Gas/Vapour	Liquid	Solid	Continuity – Continuous	Sampled	Reference section
thermal conductivity	●			●		6.2
combustibility	●			●		6.2, 6.7
chromatography	●	●			●	6.2
mass spectrometry	●	●	●	●	●	6.3
UV photometry	●	●		●		6.4
visible photometry	●	●		●		6.4
IR photometry	●	●		●		6.5
colorimetry		●	●	●		6.6
turbidimetry		●	●	●		6.6
paramagnetism	●			●		6.7
refractometry		●		●		6.8
density		●		●		5.7
pH		●		●		5.8
humidity	●			●		5.9
moisture content			●	●		5.10
radioactivity		●	●	●		6.9

6.2.1 Gas chromatographs

These are the most widely used types of on-stream analyser. They are capable of analysing up to one hundred or more individual components in a gas or vapour stream; they can separate complex mixtures into their components and measure the concentration of these components. Gas chromatographs are used on such processes as catalytic cracking, distillation, ammonia production, sulphur recovery, etc. The following factors are relevant:

- one must know beforehand the components to be expected in the sample and (approximately) their relative concentration
- all components must be volatile
- concentrations ranging from a few parts per million to approximately 60% can be measured
- variables such as flowrate and temperature must be regulated carefully, because they will affect the measurements
- vapours formed from liquids having boiling points up to 450°C may be analysed.

Figure 6.1 is a symbolic representation of a chromatograph. A minute sample (typically in the order of 0.2 μl) of the test vapour/gas — called the 'mobile phase' — is injected, possibly after being pre-heated, into a stream of carrier gas: this gas is controlled to have a constant flowrate and, typically, is either helium or nitrogen. If the test medium is in the liquid phase, it must be vaporised beforehand, using a small flash heater.

The vapour/gas mixture is passed through a packed column, which typically takes the form of a coil of stainless steel, teflon or glass tube packed with appropriate adsorbing or absorbing material. A wide range of

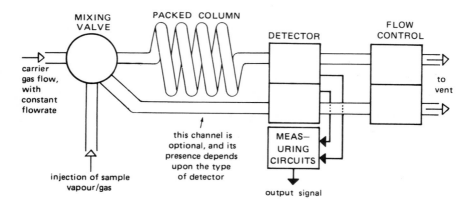

Fig 6.1 Symbolic representation of a gas chromatograph, showing the principal detection channel and optional reference channel

suitable packing materials — called the 'stationary phase' — is available: these can take the form of solid particles, sometimes coated with non-migrating liquids, or coated meshes. A typical column would have a length of 1 to 4 metres and a diameter of 3 to 6 mm. However, there are instances of columns with lengths up to 100 metres and diameters as little as 250 to 500 μm; such columns are unfilled, and rely on the adsorption properties of the inner wall of the column to function. Normally, the column will reside in a temperature-controlled oven, and in some cases the oven temperature may be pre-programmed to vary in a specific manner.

The time taken for any component of the original test sample to emerge from the far end of the column is known as the 'retention time'. Because the nature of the components in the sample is known beforehand, in qualitative terms, the column details are appropriately selected so that the retention times of the set of components in the sample are well separated in value.

The rates at which individual components of the sample flow through the column depend upon a combination of relative molecular weight and relative affinity for the column packing material. Each component is identifiable by its position in the time-based sequence as it emerges from the column and enters the detector.

The time required to perform an analysis with a single column typically ranges from 1 to 20 minutes, including any time required for subsequent flushing, so that the corresponding sample rates will range from 1 per minute to 3 per hour. However, high-speed chromatographs employing up to ten columns, each handling (typically) fifteen different components, have been produced, and these can perform an analysis in as little as 10 seconds, yielding a rate of 6 samples per minute: such an instrument requires a computer to control the complex sequencing required by the mixing valves and detectors.

The techniques which have been used for detectors for gas chromatographs include the following:

- thermal conductivity
- electron capture
- helium ionisation
- ultrasonic
- microcoulometry
- dielectric

- flame ionisation
- flame photometry (see Section 6.4)
- gas density balance
- argon ionisation
- thermionic
- infra-red (see Section 6.5).

Thermal conductivity and flame ionisation detectors are by far the most commonly used, and these will be described in Sections 6.2.3 and 6.2.4 respectively. Figure 6.2 shows typical variations in the output signal from the detector, when the sample contains four measurable components. The relative concentration of a component in the sample is indicated, in quantitative terms, by the area enclosed by the associated peak on the trace, measured above the datum line. Thus, it is necessary to compute the

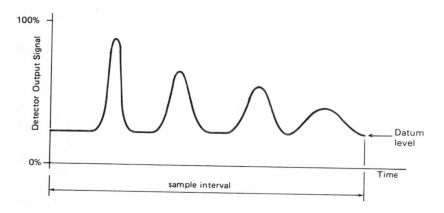

Fig 6.2   Example of the variation in detector output signal, to a base of time, typically produced by a gas chromatograph

integral of the detector output signal, after first subtracting any quiescent signal level, in order to generate the required data.

6.2.2   Liquid chromatographs

Where a liquid sample cannot be vaporised because of excessively high boiling point, it is possible sometimes to pass the liquid directly through the column. Such liquid chromatographs are similar in design and operation to gas chromatographs and are usually used with liquid samples having temperatures up to 200°C. They are also suitable for samples which are unstable or corrosive at high temperatures.

Columns usually have a short length, employ solid packing material, and operate at high pressures (up to 20 MPa). Typically, the carrier is a water/methanol mixture, and the proportion of methanol may be programmed. A test sample typically has a volume in the region of 10 μm.

The choice of alternative packing materials for liquid chromatographs is presently rather limited, with the result that cycle times between 15 and 30 minutes are typical

The detectors most commonly used with liquid chromatographs employ the following alternative techniques:

- flame ionisation (see Section 6.2.4)
- UV and visible light photometry (see Section 6.4)
- refractive index detection (see Section 6.8)
- fluorescence detection
- spray impact detection
- flame emission detection
- atomic absorption detection

- detection of electrochemical properties
- detection of chemically reactive properties.

6.2.3   Thermal conductivity detector

The temperature of a heated electrical filament, exposed to an atmosphere of a gas or vapour being measured, will depend upon the heat transferred from the filament to the gas and subsequently conducted away by the gas flow. Figure 6.3 shows a typical Wheatstone bridge network of four heated filaments (or thermistors), mounted in pairs within two chambers enclosed by a common temperature-controlled metal heat sink. The change in temperature and consequent change in resistance of the test chamber filaments will depend upon the prevailing composition of the test gas, and this will be reproduced as a change in the DC output voltage from the bridge.

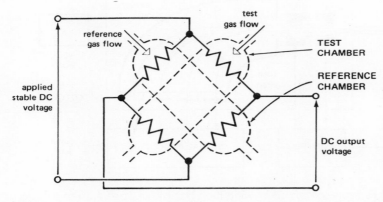

Fig 6.3   Typical Wheatstone bridge network of heated resistance elements in the two chambers of a thermal conductivity detector

6.2.4   Flame ionisation detector

The sample flow is passed into a burner assembly where it is mixed with hydrogen gas. The mixture is ignited by a heated platinum wire and is permitted to burn in an atmosphere of air or oxygen which is diffused around the burner. The temperature of the flame is such that ionisation of the sample occurs, and this results in a measurable electrical potential difference between the burner and an electrode located above the burner. This potential difference can be shown to be proportional to the number of carbon atoms currently present in the flame.

6.3   Mass spectrometers

Mass spectrometers are comparable to gas chromatographs, in terms of general performance, but the former are both more expensive and faster.

To function, the most sophisticated spectrometers use a computer which should preferably be dedicated to the one task; the simpler spectrometers will require some basic digital hardware for analysis and read-out. Mass spectrometers are used principally to determine the concentration of elemental components, and discrimination between the same type of ion from different compounds is possible, with special conditioning or signal processing. Instruments are reported to be capable of measuring the concentrations of ten to twelve components in less than two seconds, although sample preconditioning may add to this timespan; however, the majority of instruments will measure fewer components and will require a longer processing time. Figure 6.4 illustrates the principle of operation.

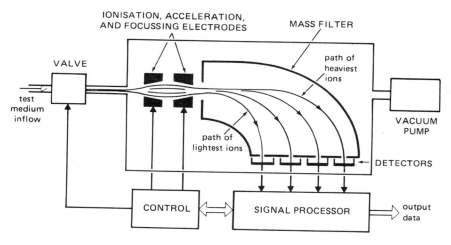

Fig 6.4 Symbolic representation showing the principle of operation of a mass spectrometer

Samples of test vapour are drawn in under suction established by the vacuum pump, at vacuum pressures as low as 10^{-4} Pa in some installations. In the case of test media which are either solids or liquids, the sample is first vaporised. In some cases, the input is derived from a chromatograph. Each input sample is ionised by one of a number of alternative means (for example, by bombardment by an electron beam), and the ion beam produced is accelerated and focussed by a suitable electric field system. The high-speed beam enters a mass filter, in which an electrostatic field system imparts a uniform kinetic energy to the ions and a magnetic field system causes the ions to deviate from a straight path. The trajectory taken by each ion depends upon its mass/charge ratio, so that each type of ion will follow a different trajectory. In the arrangement of Fig. 6.4, the detectors are ion capture devices, such as electron multipliers, located behind narrow slits: they are capable of sensing concentrations ranging from several parts per billion to several tens of percentage.

In alternative arrangements, a single detector is used and the strength of the magnetic field in the mass filter is swept as a function of time, so that the time taken for an ion to reach the detector is proportional to its mass. A typical scan takes 0.2 to 0.5 second, so that scans may be repeated at frequent intervals, to build up a series of 'fingerprints', each representing the composition of the input medium at the current moment in time. Typically, the output signal from the detector is converted to digital data at, say, 5000 conversions/second; 25 conversions may represent a Gaussian distribution of the incidence of a particular type of ion, and 500 such distributions may occur in a particular scan, corresponding to one fingerprint. Thus, high-speed signal conversion and computation are required for this type of instrument.

Alternative methods of ionisation use the following means:

- electron bombardment using heated filaments
- spark generation
- optical techniques
- chemical techniques.

Mass filters can involve the following:

- magnetic fields
- magnetic and electric fields in tandem
- sets of four charged parallel rods which produce fields having superimposed constant and time-varying components
- a series of grids polarised alternately at high frequencies.

Materials admitted to the ionisation chamber must be gases or vapours free from contaminating particles and, in the case of sophisticated spectrometers, held within narrow temperature and pressure limits. Samples may require pretreatment.

Typical applications occur in natural gas production, coal gasification, PVC manufacture, sewage treatment, and environmental control.

6.4   Ultra-violet and visible light photometers (spectrophotometers)

With spectrophotometers, analysis is achieved by passing radiation, usually from a broadband source with wavelengths in the 0.1 to 0.4 μm range, through the test medium, which may be a gas, vapour, or liquid. The medium is then identified by means of the variation in transmitted radiation as a function of wavelength. Because the absorption bands are broad and limited in number, only a restricted range of substances can be analysed in this manner: organic compounds such as aromatics, diolefins, ketones, and aldehydes may be measured to within a few parts per million; inorganic gases and vapours such as ozone, chlorine, and mercury vapour may also be measured. There are five alternative configurations, involving various orientations of radiation sources, radiation detectors, filters, beam splitters, and the test cell. These types are referred to as:

- opposed beam
- split beam
- dual beam with single detector
- dual beam with dual detector
- flicker.

Figure 6.5 shows the construction of a typical opposed beam photometer. In this, the function of the reference channel is to cause compensation for variations in the spectrum of the radiation source. This source, together with the filters (which isolate specific wavelengths), will be chosen on the basis of the chemical for which analysis is sought. Occasionally, the

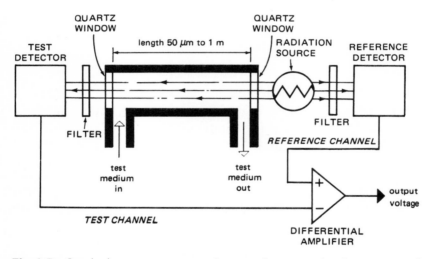

Fig 6.5 Symbolic representation showing the principle of operation of an opposed beam photometer

radiation source may be pulsed at a specific frequency; also, a sequence of filters may be rotated through the beams, in order to extend the range of the instrument. The calibration of the instrument is reasonably linear and the 90% response time is in the order of 1 second, typically.

Alternative radiation sources include:

- hydrogen discharge lamp (broadband)
- tungsten lamp (broadband)
- tunsten-iodine lamp (broadband)

- mercury vapour lamp (0.2537 μm wavelength)
- deuterium lamp (narrow band UV)
- tunable diode laser (narrow band).

Alternative radiation detectors include:

- vacuum phototube
- photomultiplier
- photo-voltaic (solar) cell
- photodiode
- phototransistor.

6.5 Non-dispersive infra-red (NDIR) analysers

These analysers function along similar principles to UV and visible light photometers, but operate in the 1 to 2.5 μm wavelength range. Usually, they are used to determine the concentration of only one component, although exceptions have been cited which can analyse up to nineteen different components.

NDIR analysers are suitable for a range of gases and liquids exhibiting good infra-red absorption: typical examples are carbon dioxide, carbon monoxide, ethylene, isobutane, etc. Specifically excluded are elemental diatomic gases such as oxygen, hydrogen, nitrogen, chlorine, etc., and inert gases, such as helium and argon.

These analysers operate on the principle that radiation energy is absorbed and converted to kinetic energy by those molecules which respond to the wavelengths of interest: it is therefore necessary to find a wavelength at which only the component of interest responds. However, overlapping effects due to other components can sometimes be accommodated, by means of appropriate compensation and filtering.

The absorption bands in the IR region tend to be narrow, so that NDIR analysers exhibit good selectivity; however, they are less sensitive than spectrophotometers. Ninety per cent response times range from 0.5 to 10 seconds.

The radiation source must be capable of radiating energy over a wide band of infra-red wavelengths and, for this reason, glass envelopes are not practicable. Alternative sources include:

- a coil or helix of metal alloy wire heated, by electric current, to the 400 to 900°C temperature range
- a rod of silicon carbide heated, by direct current, to the 1100 to 1370°C temperature range
- a rod of zirconium oxide heated, by direct current, to 1930°C
- a tunable diode laser.

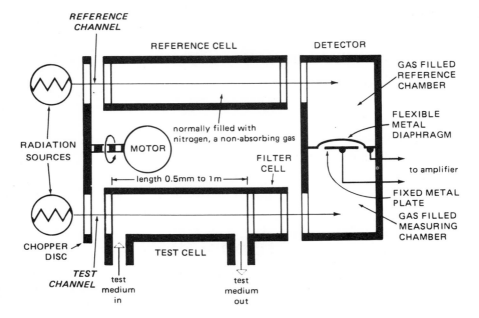

Fig 6.6   Symbolic representation showing the principle of operation of a typical NDIR analyser

Radiation detectors can be divided into three categories:

- thermal — thermopiles (thermocouple stacks) and bolometers (large thermistors)
- pneumatic — microphones and diaphragms
- photosensitive — phototubes, photomultipliers, photo-voltaic cells, photodiodes, and phototransistors.

Figure 6.6 shows a typical NDIR analyser. The gas used to fill the detector chambers is chosen to absorb energy at the wavelength of interest, and therefore is usually the same gas as the component of interest in the test medium: this detector gas absorbs energy and expands as a result. Unequal energy in the two chambers results in displacement of the metallic diaphragm. This diaphragm and the fixed metal plate together act as a capacitance microphone, which is connected into an amplifier. The filter cell is filled with selected gases, to remove energy at specific wavelengths from the test channel beam, in order to inhibit overlap: it is filled, therefore, with the component responsible for the overlap.

6.6   Colorimeters and turbidimeters

Colorimeters and turbidimeters measure the intensity of transmitted or reflected visible light. The test medium may be a gas, vapour, liquid, or solid, depending upon the construction of the instrument. Colorimeters are concerned with the precise measurement of specific colours and are used for the control of dyeing processes, the detection of parasitic substances (having specific colours) in stream flows, the measurement of chemical reactions in terms of the colours generated by the reaction processes, etc. Turbidimeters are concerned with the precise measurement of the clarity or opacity of the test medium. There are two categories of instrument: *spectrophotometric* types and *tristimulus* types, of which the latter are the more accurate.

6.6.1 Spectrophotometric types of colorimeter

These instruments are constructed along the lines of visible light photometers, as described in Section 6.4. Interference filters are selected for specific wavelengths in the visible band. Similarly, the radiation detectors are selected for operation in the visible region. Typically, the test medium would be compared with a one-dimensional colour standard reference, which is normally chosen from a series of specific concentrations of platinum-cobalt solutions. The output from the instrument will indicate the relative intensity of colour in the test medium, at the selected wavelengths.

Alternatively, if the test medium is a sample held for analysis for a specific interval of time, graduated gratings may be used to scan a series of optical wavelengths, in order to identify the colour of the sample.

6.6.2 Tristimulus types of colorimeter

These instruments measure transmitted or reflected light on the basis of comparison with specific colour standards. The standards usually used have spectral responses in the blue, green, and amber bands, as indicated in Fig. 6.7.

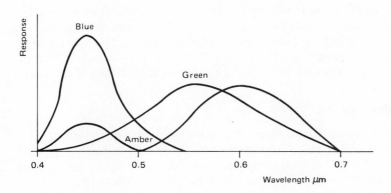

Fig 6.7 Spectral responses of blue, green and amber colour standards

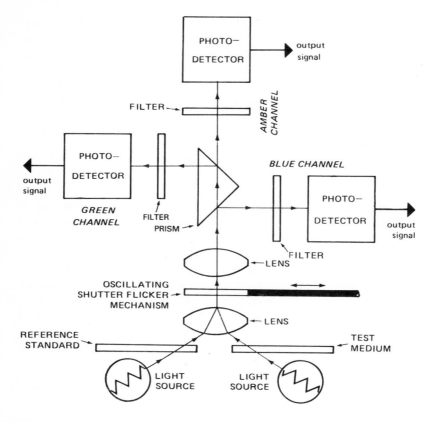

Fig 6.8   Symbolic representation showing the principle of operation of a typical tristimulus colorimeter

The instrument depicted in Fig. 6.8 produces a rapid, sequential, comparison between the test medium and the reference standard, on the basis of the measurements made by the three different colour channels. The instrument measures directly the percentage difference in light energy between the test medium and the reference. Because the measurement is in terms of ratio determination, the flicker technique reduces the effect of drift caused by long-term variations in the light source and photodetectors.

6.6.3   Turbidimeters

Turbidimeters are constructed along the same lines as colorimeters. The light sources and filters in colorimeters are chosen to enhance specific wavelengths of interest; in turbidimeters, however, the light sources and filters are chosen to be non-specific: that is, no particular wavelength should be favoured. Thus, the photodetector used will be sensing the intensity of broadband light received, and this will be a measure of the

clarity or opacity of the test medium. A common application for turbidimeters is in smoke level detection in the smoke stacks of power stations
and other plant.

6.7   Oxygen analysers

There are many applications in which it is required to sense oxygen levels.
One example is the measurement of oxygen level in smoke stacks, because
this represents a good indication of boiler combustion efficiency. Another
example is the measurement of oxygen level in water, because this
represents a good indication of the level of possible pollution.

Most types of oxygen analyser use either the combustible property or
the 'paramagnetic' property of oxygen, but there are a few based upon
other principles. All types have a 90% response time no greater than one
minute. The paramagnetic property of oxygen refers to the fact that
oxygen molecules are attracted by a magnetic field; other gases, with the
exception of a few oxides of nitrogen, are either repelled by magnetic
fields, and are referred to as 'diamagnetic', or are unaffected.

6.7.1   Combustion-type oxygen analysers

With these analysers, the oxygen content in a gaseous test sample is used to
oxidise a fuel, and the heat so generated from subsequent combustion is
measured. In the configuration of Fig. 6.9, the test sample and a measure
of fuel are mixed in the mixing chamber, and combustion in the measuring
cell is stimulated by a catalytic coating on the measuring filament. The two
filaments serve to sense temperature rise, being connected as two adjacent
arms in a Wheatstone bridge network. The function of the (uncoated)

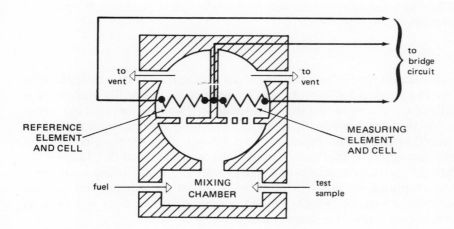

Fig 6.9   Arrangement of a typical combustion type of oxygen analyser

reference filament is to compensate for variations in temperature and thermal conductivity of the test sample.

The bridge network senses the relative changes in filament resistance, which will be related to combustion temperature and hence to oxygen level in the test sample.

6.7.2   Paramagnetic oxygen analysers

Paramagnetic analysers are distinctive from the combustion types of analyser in that the test medium is not destroyed in the measurement process. Paramagnetic analysers can be subdivided into two categories: the deflection type, which requires the paramagnetic property to be constant during the measurement, and the thermal type, which depends upon the decrease in paramagnetic effect which results from an increase in the temperature of the oxygen content in the test medium.

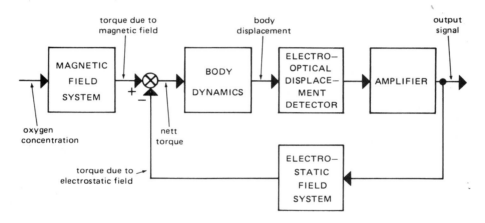

Fig 6.10   Block diagram representation of a deflection type of paramagnetic oxygen analyser

The deflection type of analyser operates on the torque balance principle described in Section 4.4.1, and seeks to balance magnetically and electrostatically derived torques arranged to act upon a body suspended in the gas stream, as indicated in Fig. 6.10. Typically, the body is a small dumbbell suspended, by a filament, between the poles of a magnet; the electrostatic field is established by electrodes sited close to one end of the dumbbell.

When the instrument settles to a steady state, a torque balance will exist and the amplifier output will be a measure of the torque due to the magnetic field system, which will be a known function of the oxygen concentration around the magnet and dumbbell.

Figure 6.11 shows the construction of a thermal type of analyser. The oxygen component in the gas stream passing through the glass vessel is attracted by the magnetic field, and tends to pass down the central glass tube, where it is heated by the heater windings. These windings form two adjacent limbs of a Wheatstone bridge network, and the change in resistance sensed will be a function of the conduction of heat by the oxygen flow. The oxygen loses much of its paramagnetic property as it becomes heated, and convection causes more oxygen to be drawn in past the pole face. The rate at which oxygen is able to flow past the pole face will be a measure of its concentration, and this will be indicated by the output from the bridge network: the left-hand winding tends to be cooled by the entry of fresh oxygen, whilst the right-hand winding tends to be heated by the heat drawn from the left-hand winding by the oxygen flow.

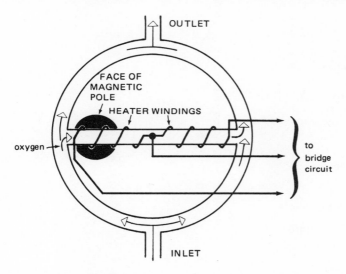

Fig 6.11 Configuration of a thermal type of paramagnetic oxygen analyser

6.7.3 Dissolved-oxygen analysers

The most commonly used analysers for measuring the concentration of dissolved oxygen in liquid streams are the *polarographic* and *galvanic* types. Note, however, that analysers normally used for measuring the oxygen content in gas flows may be adapted to the measurement of dissolved oxygen, if the oxygen is removed beforehand from the liquid stream, using appropriate means; also, polarographic and galvanic analysers may be used for the measurement of gaseous oxygen.

Figure 6.12 shows a cross-section through a typical polarographic probe. The probe incorporates two noble metal electrodes, which are separated from each other by an electrolyte of potassium chloride solution or gel. The probe face is covered by a permeable Teflon membrane, which separates the gold cathode from the test medium. A constant voltage of approximately 0.8 V DC is applied between the electrodes. The oxygen in the test medium permeates the membrane and diffuses to the cathode, where it is reduced, causing a flow of ions to the anode. The current flow around the circuit formed by the voltage source, electrodes, and electrolyte is proportional to the oxygen concentration and is amplified by a suitably configured instrumentation amplifier. The calibration is sensitive to temperature, and thermistors embedded in the probe are used to generate a signal which is applied to compensate for temperature variations.

Figure 6.13 shows a cross-section through a typical galvanic cell, through which the test medium flows, making contact with two electrodes, one of noble metal and the other of base metal alloy. The two electrodes are

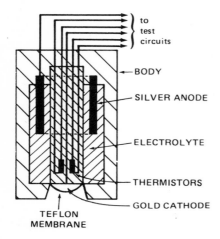

Fig 6.12   Cross-section through a polarographic dissolved-oxygen analyser probe

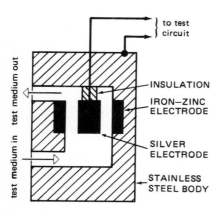

Fig 6.13   Cross-section through a galvanic dissolved-oxygen analyser cell

polarised by an externally connected DC voltage source. The liquid stream serves as the electrolyte and the dissolved oxygen becomes ionised, resulting in a drift of ions between the electrodes. The current flow is measured by a suitably configured instrumentation amplifier, and is proportional to the oxygen concentration. The cell is sensitive to temperature and pressure variations, which must be either controlled or measured and compensated. The liquid medium must have a minimum conductivity for the cell to function, so that a suitable chemical might need to be added to the stream, upstream of the cell, in order to achieve this requirement.

6.8   Refractometers

Refractometers are used for measuring the relative concentrations in liquids composed of two principal components, and the measurement will be degraded by the presence of trace impurities.

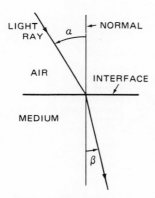

Fig 6.14   Diagram for the definition of refractive index

Figure 6.14 shows how the refractive index n is defined, on the basis of the refraction of a light ray at the interface between air and a test medium:

$$n \stackrel{\Delta}{=} \frac{\sin \alpha}{\sin \beta}$$

Figure 6.15 shows alternative paths for light rays impinging on the interface between two media, with the light source being sited on the side occupied by the medium having the larger refractive index. Total internal reflection just occurs when the angle of incidence is equal to the critical angle ϕ_{crit} as shown, and this value is given by $\phi_{crit} = \sin^{-1}(n_B/n_A)$.

One type of refractometer, which is used with clean translucent liquids, is based upon changes in refraction angle. The other type, which is used with clean or turbid liquids or slurries, is based upon changes in critical

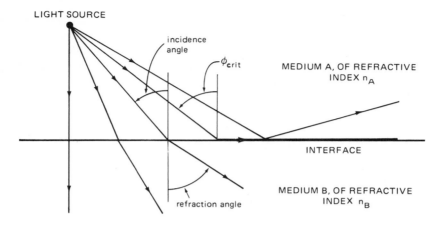

Fig 6.15   Refraction and reflection of light rays having different angles of incidence at the interface between two media

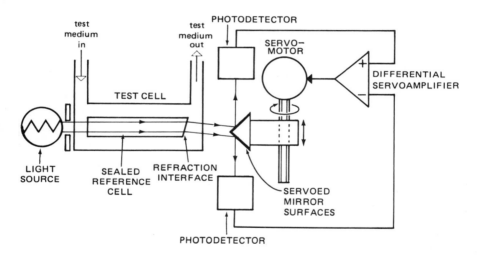

Fig 6.16   Symbolic representation showing the principle of operation of a typical refraction angle type of refractometer

angle. In each case, medium A would be a fixed reference and medium B would be the test medium.

Figure 6.16 shows a typical arrangement for a refraction angle type of refractometer, which operates on an illumination balance principle. The reference cell is filled with a liquid having a refractive index which is approximately mid-scale for the test medium. The servomotor drives the beam splitter until the two photodetectors sense equal incident illumination. In the steady state the position of the beam splitter will be a measure of the refraction angle of the beam, which will be representative of the composition of the test medium. The typical 90% response time for refractometers is in the 10- to 30-second range.

6.9 Radiation techniques

Where radioactivity is to be used for measurement purposes, the choice of radiation source will depend upon the type of radiation to be used, which in turn will depend upon the test medium and its surroundings. The choice of detector for sensing the intensity of radiation transmitted through the test medium will also depend upon the type of radiation, in addition to other factors.

6.9.1 Radiation and radioactive sources

In some cases, radioactivity can be defined in terms of the emission of specific subatomic particles. In other cases, it is convenient to regard electromagnetic radiation as a stream of small packets (quanta) of energy, travelling at the velocity of light.

The α particle is the nucleus of the Helium-4 atom, and is therefore a body consisting of two protons and two neutrons, bound together. It is characteristic of very heavy elements; the energy level associated with it can occur within the 2 to 9 MeV range, but the level lies within a very narrow band of MeV for any particular radioisotope. The α particle will be stopped completely even by very thin material, so that it has a very limited industrial use.

The β particle is the electron ejected from an atomic nucleus, when an excess neutron decays into a proton plus an ejected electron. The energy level associated with the β particle can cover a wide MeV band for each radioisotope, with the maximum level lying between 18 keV and 3.6 MeV, typically. The β particle can pass through steel up to 3 mm in thickness.

γ radiation is an electromagnetic radiation emitted when an atomic nucleus in an excited state reverts to a more stable state. The radiation can occur with a number of simultaneous specific quantum energy levels lying between 100 keV and 3 MeV. γ radiation can pass through steel more than 30 cm thick and concrete more than 100 cm thick. The intensity of a beam of γ radiation is diminished by solid matter in inverse proportion to the distance travelled through that matter: the more dense the material, the more effective it is in attenuating the radiation.

X radiation is an electromagnetic radiation emitted when an electron transfers from a higher to a lower atomic energy state. X radiation is similar to γ radiation but has much lower energy levels associated with it.

Other types of electromagnetic radiation are emitted by certain radioisotopes, but normally these other types are not significant as far as industrial measurements are concerned.

Artificial radioisotopes are produced by exposing target material to bombardment by high-velocity particle streams, in cyclotrons and nuclear reactors. Over 700 types of artificial radioisotope have been made, emitting a variety of radiation (α, β, γ, X, etc.) and having half lives ranging from 2 minutes to 5×10^{10} years. Certain radioisotopes emit only one type of radiation, whilst others emit several types simultaneously. For any given radioisotope, the energies of the emitted radiations are constant and specific to that radioisotope.

The extent of the energy loss occurring when a test medium is exposed to radiation depends upon the type of radiation. Consequently, the choice of a radioisotope to be used as the radiation source in any particular application will be based upon the properties and physical dimensions of the test medium.

6.9.2   Radiation detectors

In the industrial applications of radiation techniques considered in this volume, the type of radiation involved and the energy levels associated with it will be specific for the radioisotope chosen to be the radiation source: this information therefore will represent known data. The function of the detector therefore can usually be confined to the measurement of the quantity of radiation arriving at the detector; in addition, some applications may necessitate the measurement of the spatial distribution of this radiation.

The range of radiation detectors used industrially includes *ionisation chambers*, *proportional counters*, *Geiger Muller counters*, *scintillation counters*, and *solid state detectors*, together with other less common types.

Ionisation chambers

Figure 6.17 shows a cross-section through a typical ionisation chamber. In industrial applications, it would be used most widely for the detection of β and γ radiation: in the former case, the β particles would enter through a quartz or glass window inserted in the wall of the chamber; in the latter case, γ radiation would be able to pass through the metal wall without difficulty, so that a window would not be required. Typically, the gas filling would be argon, and may be highly pressurised for the detection of γ radiation. The applied DC voltage will have a value of several hundred volts. Radiation entering the chamber causes ionisation of the gas molecules, and positive ions will migrate to the cathode whilst electrons will migrate to the anode.

The resulting current circulating in the circuit is detected by a suitably configured instrumentation amplifier. Each ionising event will generate a minute current pulse: these pulses may either be counted individually, to yield a total count size, or integrated to produce a measure of the frequency with which the events occur.

Proportional counters

The proportional counter is a special form of ionisation chamber operated at rather higher levels of applied DC voltage. At such levels, secondary ionisation occurs, with the result that the chamber can establish

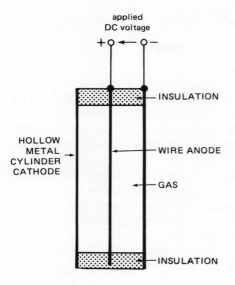

Fig 6.17   Cross-section through a typical ionisation chamber

(internally) charge amplification factors as high as 10^4. The filling may be argon, hydrogen, methane, or some other gas, at various alternative pressures. This type of detector is always used for pulse counting, and is particularly suitable for high pulse frequencies and for distinguishing between different types of radiation; generally, it is used for α and β particle detection.

Geiger Muller counters

The Geiger Muller counter is another special form of ionisation chamber, but it is excited by voltage levels well in excess of 1000 V. At these levels, an avalanche breakdown occurs, generating charge amplification factors as high as 10^8. Because of the high signal levels generated, external amplification is rarely necessary. Typically, the filling is a mixture of argon and an organic vapour. This type of detector is always used for pulse counting, but it is unsuitable for very high pulse frequencies; generally, it is used for α and β particle detection.

Scintillation counters

A range of materials, which may be crystals, liquids, gases, plastics, or certain types of glass, behave as phosphors: that is, they emit photons when impacted by quanta of radiation. A scintillation counter uses a mass of one of these phosphor materials, and the light emitted as a result of exposure to radiation is converted to a current and amplified, using a photomultiplier, which can achieve current gains as high as 10^6. This type of detector may be used for sensing either pulse count or pulse height, so that it is able to distinguish between different radiation energy levels. It can operate at very high frequencies and is suitable for the detection of α, β and γ radiation.

Solid state detectors

A *semiconductor radiation detector* is based upon a silicon or germanium p-n junction, across which a small DC voltage is applied. The incidence of radiation upon the junction causes an increase in hole-electron pairs, so that a pulse of current is generated in the external circuit, which can be configured to count such pulses. These detectors are suitable for detecting α and β particles.

An alternative type is the *photoconductive detector*, which is based upon a crystal of high purity. The incidence of radiation causes an increase in the conductivity of the crystal, so that a current pulse is generated in the external circuit, which is configured to count such pulses. These detectors are suitable for the detection of β and γ radiation.

7

Recording Instruments

7.1 Introduction

Recording instruments are frequently to be found operating in close conjunction with feedback control systems; indeed, many such instruments were in common use long before feedback control was introduced on a widespread scale. Recording instruments may be used at the design and commissioning stage of a control system, to gather data so that the system may be tuned for optimum performance. They may be used with control systems in service, in order to generate a record of the performance of those systems. Additionally, instruments may be used, in their own right, to monitor and record the behaviour of plant not being automatically controlled.

This chapter begins with a brief review of the history of recording instruments, then surveys the types of recorder in common use, and finally describes in detail all of the alternative mechanisms used in these instruments.

7.2 A brief history of instrumentation recorders

Large, pneumatically activated, process controllers were introduced into industry in 1929. These were associated with pneumatically activated circular chart recorders, having 8-, 10- and 12-inch diameter charts.

Subsequently 10 inch wide rectangular chart recorders were introduced. These were also pneumatically activated and had the ability to record the variations in several pneumatic signals: see Multipoint recorders in Section 7.3.

1946 saw the introduction of combined recorder-controllers, again pneumatically activated and having 12-inch diameter charts. Subsequently, 'miniature' versions, having a 6-inch by 6 inch front panel area, were introduced: as a result of the smaller face area, the hardware extended much further behind the panel than had previously been the case. In 1958, electronic recorder-controllers became available, and at first these emulated their pneumatic counterparts. To meet this competition, a new generation of pneumatic recorder-controllers was introduced, in 1965.

To meet the requirement for greater complexity of control panel instrumentation, the trend then came for only the instrument displays to remain on the panels, whilst the associated circuitry hardware was progressively extended back behind the panels. Initially, the proportion of on-panel to behind-panel distribution was about 1:1 but over time this changed to about 1:4, with the behind-panel hardware now being mounted on separate racks. This was feasible only with electronic hardware, thus precipitating the decline of pneumatic instruments.

More recently, hardwired panel instruments have been unable to provide the concentration of instrumentation required for many modern plants, and they have progressively been superseded by CRT displays synthesising instrument displays and digital data loggers taking over the task of recorders. In addition, keyboards have superseded switches and knobs, in providing the operator with the means to interrogate and adjust instrumentation.

In larger installations, the trend has been to provide operators with control desks, either to replace or to complement wall-mounted mimic diagrams and instrument panels. Where CRTs and keyboards are provided for the operator, control desks become an essential feature of control rooms.

7.3   Types of instrumentation recorder

Recorders translate transducer signals into stylus movements across paper charts. Usually, the chart is moved at a constant speed by an electric motor (which may be either a synchronous or stepper type); occasionally, a spring-powered clockwork motor or an air fluid motor may be used.

The alternative processes for recording a trace are as follows.

- Ink pens — the ink reservoir may be directly over the stylus or it may be some distance away and connected to the stylus by a capillary tube, which may be pressurised. The stylus may be a fine metal tube or a fibre tip.

- An electrically-heated stylus running over a heat-sensitive paper.

- An electrically-energised (typically 400 V) stylus running over a voltage-sensitive paper, the surface of which breaks down at the point of contact.

- A sharp stylus cutting the surface of a wax-coated paper.

- A fine UV light beam being deflected across a photosensitive paper.

The ink pen type of recorder will have the slowest response (bandwidth in the region of 1 Hz) whilst the UV recorder will have the fastest, with bandwidths up to 20 kHz being possible.

Recorders may record only one trace, or else they may record several traces simultaneously, either superimposed or restricted to operate within non-overlapping bands along the paper. Multipoint (sampling) recorders enable multiple traces to be recorded by a single stylus, by sampling the incoming signals sequentially and recording the sample value as a point on the paper: such recorders can be used only for recording relatively slowly changing signals.

With round chart recorders, the chart diameter can be up to 12 inches (30 cm). The chart has a scale of concentric circles, with increasing values running from the inside outwards, together with timing arcs. Timing ranges can vary from 15 minutes to 28 days for the full revolution. The pen moves in an arc subtending 45 degrees, approximately, corresponding to a radial distance of approximately 4 inches (10 cm). A pointer attached to the pen arm usually sweeps across a concentric scale, attached to the instrument face and scaled in values of process variable. The instrument is mechanically simple and the charts are flat and compact. Each recorder may have from one to four pens, and some sampling versions are in use.

With strip chart recorders, the paper may be pre-folded or on a roll, and the measurement lines are straight and parallel, running down the length of the strip. The timing lines may be straight, and perpendicular to the measurement lines, or they may be circular arcs: this will depend upon the type of pen/stylus mechanism. A pointer attached to the pen/stylus arm usually sweeps across a scale, attached to the instrument face and scaled in values of process variable. Increasing values usually run from left to right, with the paper moving downwards. Each recorder may have from one to ten pens/styli, and sampling versions are in use: in the latter case, points may be in different colours or may be identified by distinctive characters placed along the edge of the paper. Digital strip chart recorders eliminate the stylus motion, and the data are represented by a series of points along the paper; alphanumeric characters may also be printed concurrently.

7.4 Instrumentation recorder mechanisms

When choosing the type of recorder mechanism most suitable for a particular application, the following factors would be relevant:

- the nature and range of the input signal
- the maximum rate of change of the input signal, in relation to the slew rate limit of the recorder
- the number of input signals to be recorded
- the accuracy required

Seven alternative types of recorder mechanism will now be described.

7.4.1 Pneumatic mechanism

With pneumatic recorders, the paper chart is driven at constant speed, using a drive involving a geared electric motor. The pen assembly is connected, through linkages, to the moving portion of one of the pressure sensors listed in Section 5.4: a Bourdon tube, bellows, diaphragm or capsule. Depending upon the linkage arrangement, the pen motion may be in the form of either a circular arc or a straight line. Such recorders are normally used for recording 3 to 15 psi pneumatic signals and, as such, would rarely feature in modern installations.

7.4.2   Moving coil mechanism

This mechanism is similar to the d'Arsonval movement used in conventional DC moving coil instruments, except that it may be rather more rugged in construction. A DC current, equal to or proportional to the signal to be recorded, is passed through a coil which is suspended, in jewelled bearings, around a solid cylinder of soft iron. A U-shaped permanent magnet completes the magnetic circuit. The current in the coil reacts with the magnetic flux and the torque thereby developed causes the coil to displace around its axis of suspension: this angular deflection is opposed by spiral springs coupled to the axis of suspension. The nett angular deflection will be proportional to the magnitude of DC current passing through the coil. Figure 7.1 shows the general arrangement. The pen will describe a circular arc.

The principal disadvantage with this type of arrangement is that precision errors can arise from the presence of static friction, especially because the actuating torque is developed directly by the input signal. The principal advantage is the relative simplicity of the mechanism.

Ultra-violet recorders use a similar mechanism, with the input current being passed through a a lightweight coil suspended by ligatures within the magnetic field passing between the poles of a U-shaped permanent magnet. A tiny mirror is cemented to the coil, and this reflects a fine beam of UV light projected from a small gas discharge lamp. This type of

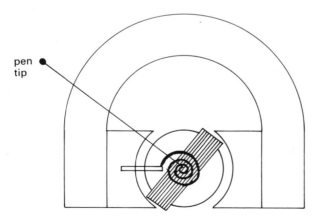

Fig 7.1   Moving coil mechanism

actuating mechanism is a form of galvanometer. The minute mass of the moving parts can result in a very high frequency bandwidth: up to many kHz in some instances.

7.4.3 Potentiometric mechanism

This mechanism involves the use of a position servosystem, with pen position feedback being generated by a potentiometer. The servosystem is similar to those described in Section 13.4. The input signal must be converted to a suitable DC voltage using, if necessary, a suitable signal converter: this voltage becomes the reference signal for the servosystem. Figure 7.2 shows the general arrangement.

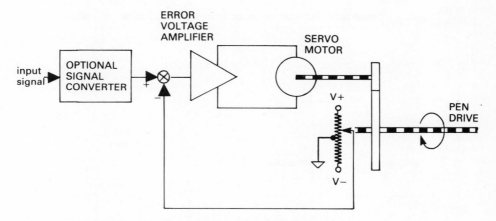

Fig 7.2   General arrangement of a potentiometric mechanism

The pen drive can be configured to yield either angular or rectilinear motion of the pen across the chart paper. The feedback potentiometer may be rotary or rectilinear. The servomotor may be either DC or AC: in the latter case, modulation of the DC error voltage will be necessary, within the error voltage amplifier.

The principal advantage with this mechanism is that the input signal source does not provide the motive power for the pen mechanism, so that high precision is possible.

7.4.4   Bridge mechanism

This mechanism also uses a position servosystem, but the reference transducer, feedback transducer, and error signal generator are combined in the form of a Wheatstone bridge. The reference transducer will usually be in the form of a variable resistance, so that typically it will be either an RTD or thermistor. Figure 7.3 shows a typical arrangement. As with

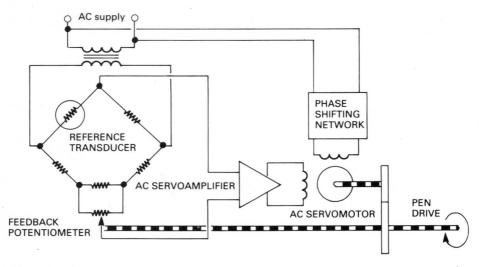

Fig 7.3   General arrangement of a bridge mechanism

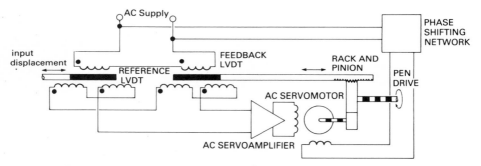

Fig 7.4   Arrangement of a typical LVDT recorder mechanism

the potentiometric mechanism, the servomotor may be AC (as shown in Fig. 7.3) or DC, and the feedback potentiometer may be rotary or rectilinear. The servomotor will continually drive the potentiometer so as to keep the Wheatstone bridge in balance, as detected by the servoamplifier. Again, the pen may have angular or rectilinear motion, and the input transducer is not required to provide the motive power for the pen mechanism.

7.4.5   LVDT mechanism

This mechanism also uses a position servosystem, but with the feedback transducer here being an LVDT. In the arrangement shown in Fig. 7.4, the instrument responds to an input displacement, which is sensed by a second LVDT, which is acting as a reference transducer. The AC servoamplifier amplifies any difference existing between the two LVDT output signals, thereby causing the servomotor to drive the core of the LVDT (and therefore also the pen) to reduce this difference towards zero value. In so doing, the displacement of the pen will become a replica of the input displacement.

7.4.6   Force balance mechanism

This mechanism also uses a negative feedback technique, but in this case it is force rather than displacement which is fed back. Figure 7.5 shows a typical arrangement. The reference force, acting on one end of the beam, is generated by the current in the coil of the electromagnet: this current is created by the aggregate of the DC power supply voltage and the DC input

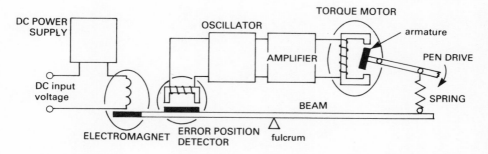

Fig 7.5   Arrangement of a typical force balance recorder mechanism

voltage. The feedback force, acting upon the other end of the beam, is generated by the torque motor and transmitted, by the spring, to the beam. Any displacement of the beam from its null position will be sensed by the electromagnetic error position detector: any change in the inductance of this device will change the tuning of the oscillator, and this causes the amplifier to develop a DC output voltage which excites the torque motor coil. As a result, the armature is attracted, returning the beam to its null position. With the beam at this position, the change in displacement of the armature shaft will be a replica of the change in the DC input voltage to the recorder mechanism.

7.4.7   Linear array mechanism

Recorders using any of the mechanisms so far described typically may contain 50 moving parts. A recent addition is the type of recorder using a linear array mechanism: this completely eliminates both the moving parts associated with the writing mechanism, and the need for ink pens, which always experience problems.

The stationary print head presents a rectilinear array of thermal styli, extending across the width of the chart paper. Normally, the heat-sensitive paper is stepped past the print head by a stepper motor drive (see Section 11.4.3), and appropriate styli are energised under the control of a micro-processor. The pattern of energisation is updated at each step of the paper drive, to reproduce analog input signals as sequences of dots; alpha-numeric characters may also be printed, for identification purposes and to indicate scales. Typical resolution is 0.25% for a 4 inch (10 cm) width, and 200 mseconds would be a typical minimum sampling interval.

8
Electric Amplifiers and Final Control Elements

8.1 Introduction

With the exception of rectilinear electromagnetic actuators ('force motors') and linear induction motors, all electrical drives generate rotary motion which, if rectilinear motion is desired, must be converted using screw-nut or rack-pinion arrangements. In the majority of cases, the drive will need to be reversible, but, in those cases where reversibility is not desired, the specifications for the drive amplifiers can be relaxed.

The general arrangement of amplifiers and motor is shown in Fig. 8.1. In terms of power requirements, the motor power rating must be matched to

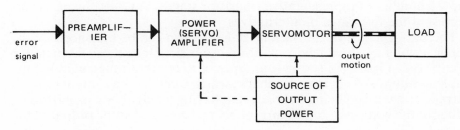

Fig 8.1 General arrangement of amplifiers and motor in an electromechanical control system

the requirements of the load. The power amplifier is required to manipulate the amount of power delivered by the motor to the load: the power rating of the power amplifier must match the power rating of the motor in those applications where the amplifier directly controls the full power delivered by the motor, whereas in other configurations the power requirements of the power amplifier may be only a small fraction of the power delivered by the motor. The power rating of the preamplifier would normally be low, since its principal function is usually signal amplification and not power amplification.

8.2 Preamplifiers

The various stages of preamplification can be called upon to implement some or all of the following tasks.

● Signal combination – that is, summation or subtraction (see Section 12.2).

● Signal (usually voltage) amplification.

● Signal conditioning – for example, noise filtering, level shifting, shaping with nonlinear static characteristics, etc.

● Signal conversion – for example, current-to-voltage conversion.

● System compensation, by the incorporation of active or passive R-L-C circuits to implement a specific transfer function.

● Impedance level changing – for example, to present a high impedance load to a transducer signal source.

Normally, the signals to be processed, using analog computing techniques, are DC voltages. In some applications, the voltages may be AC, operating at a carrier frequency which is typically either 50, 60, 400 Hz or several kHz. Preamplifiers may be designed using discrete components, in which cases the stages in the amplifier would normally be direct-coupled when DC signals are being amplified. In the case of the amplification of AC signals, the stages in the amplifier would normally be AC coupled, using either capacitors or signal transformers to transmit AC components and to block the transmission of DC components. Figures 8.2, 8.3 and 8.4 show a number of representative transistor stages, but these by no means represent the full range of alternatives.

The performance of the various stages shown can be enhanced by the application of negative feedback, for reasons similar to those discussed in Section 8.5. Figure 8.5 shows four types of network arrangement for the implementation of negative feedback, which can take the form of either a current or a voltage being fed back and having a magnitude proportional either to the voltage or to the current being developed at the output of the stage.

In modern practice, nearly all preamplifier configurations are integrated circuit operational amplifiers (IC Op. Amps), which employ stages of amplification and negative feedback arrangements of the types shown in Figs 8.2, 8.3, 8.4 and 8.5. These amplifiers exhibit the following advantages:

● high open loop voltage gain
● high frequency bandwidth
● high input impedance, especially with those having a FET input stage
● low output impedance
● low input voltage offset
● low input bias current, especially with those having a FET input stage

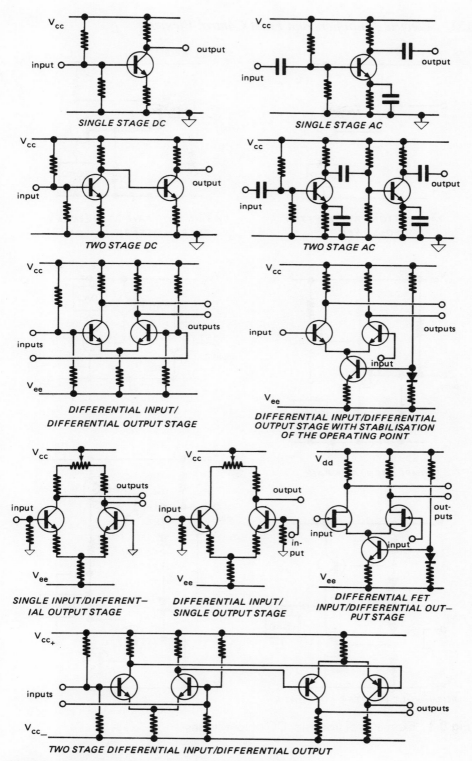

Fig 8.2 Examples of stages in transistor amplifiers

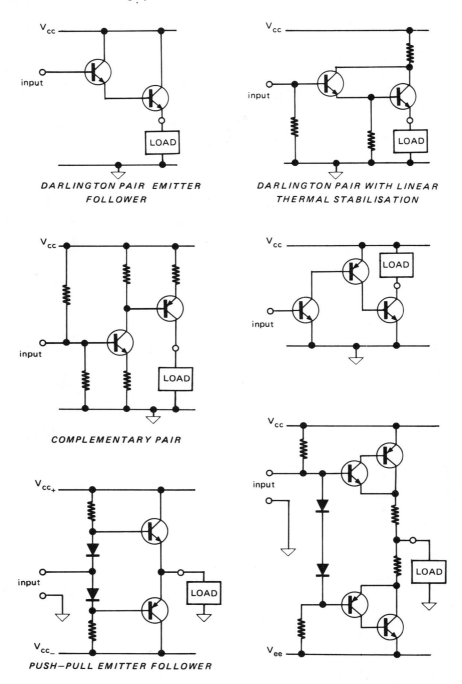

Fig 8.3   Examples of transistor DC output stages

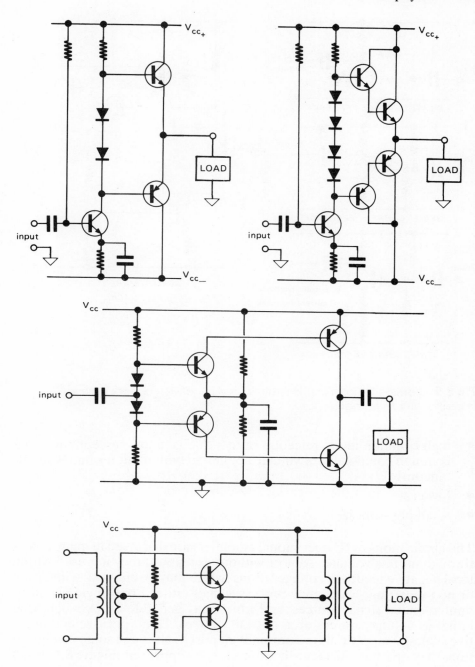

Fig 8.4   Examples of transistor AC output stages

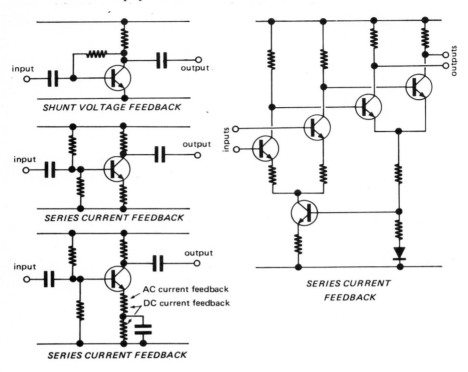

Fig 8.5 Some alternative implementations of negative feedback around transistor amplifier stages

- high common mode rejection ratio – that is, a low susceptibility to a common component of voltage applied to both input terminals simultaneously
- low cost
- small physical size.

The applications of IC operational amplifiers are discussed in many places throughout this volume. Almost without exception, they are used within local negative feedback configurations, which may be either sign inverting or non-inverting, which may process signals either from single-ended or double-ended signal sources, and which may generate outputs behaving either as voltage sources or as current sources. Dual (positive and negative) voltage supply rails are normally required, although some configurations may operate satisfactorily from a single supply rail. Figure 8.6 shows the usual symbol used to depict an operational amplifier; note, however, that in this volume this same symbol has also been used to depict a closed loop differential-input amplifier, as described in Section 12.2.3. To prevent confusion, the legend *differential amplifier* has been placed against the symbol, where this is relevant.

Figure 8.7 represents one version of a *chopper amplifier*, which uses an AC amplifier in order to amplify a DC voltage. With this, the DC input voltage is converted to a squarewave of comparable amplitude, using a solid state switching 'chopper' circuit, which acts as half wave squarewave modulator. The alternating component of this squarewave is amplified by the AC amplifier and subsequently demodulated by a phase-sensitive demodulator and smoothed, to eliminate switching spikes. The DC output voltage is an amplified version of the DC input voltage, and the chopper amplifier exhibits very high voltage gain stability and very low input voltage and current offsets; however, its bandwidth is limited by the frequency of the oscillator output.

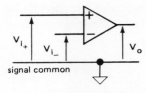

$$V_o \cong A \, (V_{i_+} - V_{i_-})$$

where A = open loop voltage gain

Fig 8.6   Symbolic representation of an operational amplifier

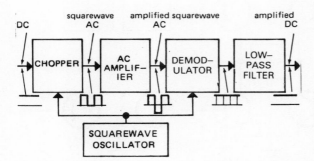

Fig 8.7   Block diagram representation of the stages within a chopper amplifier

8.3   Transistor power amplifiers

Transistor power amplifiers normally use direct-coupled transistor stages and are biassed to operate in the Class AB mode. Their principal function is to amplify power levels, so that the voltage amplification may be small or even unity. Negative feedback is often incorporated into the amplifier. Typically, the amplifier output would drive a winding, or windings, on a servomotor.

Typical considerations to be made when developing a power amplifier would be:

- the voltage, current and power requirements of the load
- the availability of suitable DC power supplies

- the requirement (or otherwise) for reversibility of the output voltage
- the nature of the load – whether single-ended, double-ended, centre-tapped or split
- the probable requirement for a push-pull output stage, which can be used to minimise crossover distortion.

Figure 8.8 gives examples, simplified, of some alternative configurations for output stages and loads: all can develop a reversible output voltage. For non-reversing output requirements, simpler output stages may be used.

Some monolithic power amplifiers are manufactured and these can develop several tens of watts of output power. They can be cascaded with an IC operational amplifier, and then negative feedback would typically be applied around the combination. Figure 8.9 shows two examples, both of

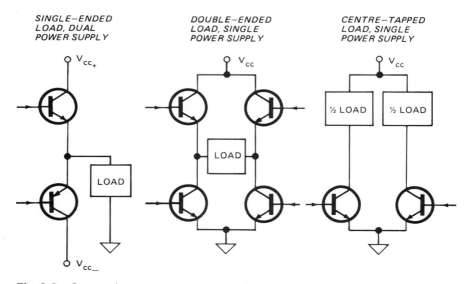

Fig 8.8 Some alternative transistor power output stages (simplified) with different load and power supply configurations

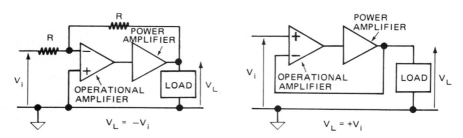

Fig 8.9 Alternative negative feedback configurations using an operational amplifier and a monolithic power amplifier

which can drive only a single-ended load. Alternatively, the output stages of Fig. 8.3 could be used as the power amplifiers in the Fig. 8.9 configurations.

Figure 8.10 shows a typical discrete component servoamplifier, which uses an IC operational amplifier as an integral preamplifier and which is suitable for amplifying either DC or AC voltages. It incorporates negative feedback (in fact, a feedback current proportional to a fraction of the output voltage) which has been applied around the complete amplifier.

The power amplifier designed for Class AB operation can generate an output waveform which is a faithful (amplified) reproduction of the input waveform up to signal frequencies in the region of 10 kHz. Although the

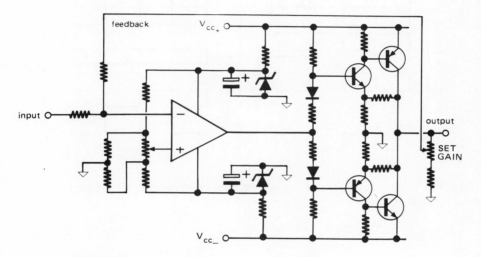

Fig 8.10 Circuit for a typical servoamplifier using discrete components and an operational amplifier

output from a Class AB amplifier essentially is ideal, the method of amplification results in substantial internal power dissipation.

Switching transistor amplifiers employ high-speed switching transistors to provide a proportional output. This type of amplifier has at least two transistors connected to a positive and negative voltage generated by a high power internal power supply. In contrast to the Class AB amplifier, the output transistors in a switching amplifier are in either a totally conducting or a non-conducting state. Power control is obtained by modulating the pulse width duty cycle in accordance with the magnitude of the input control signal, so that the average value of the output voltage is proportional to the input signal. Figure 8.11 shows an example of a switching amplifier, using field effect power transistors. The switching frequency is typically between 1 and 25 kHz. The relatively high switching frequency provides excellent reproduction, up to about one third of the

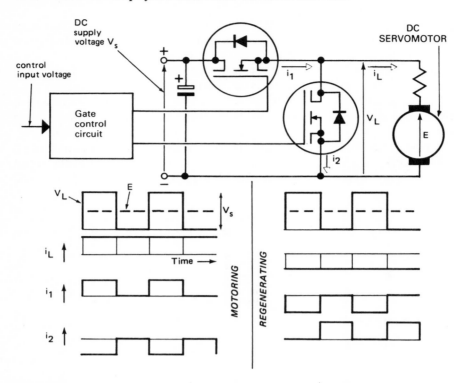

Fig 8.11   Circuit diagram and waveforms for a switching amplifier using field effect power transistors

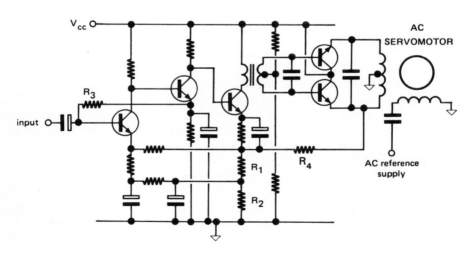

Fig 8.12   Circuit diagram of a typical AC servoamplifier

switching frequency. The switching transistor amplifier offers excellent power efficiency because the output transistors are either totally on or off.

Figure 8.12 shows a typical AC servoamplifier, capable of controlling a small two-phase servomotor of the type described in Section 8.6.1: it incorporates several feedback paths, involving voltages developed across resistors R_1 and R_2 and currents flowing through resistors R_3 and R_4.

8.4 High-power power output stages

Where a power transistor output stage cannot meet the power requirements of the load, it normally becomes necessary to employ either a DC generator, a Triac network, or an SCR network. Note, however, that power MOSFETs are being manufactured with increasingly high power ratings and that eventually networks using these devices may take over many of the roles for which Triac and SCR networks are currently used. The MOSFET exhibits the advantages of very fast switching, low ON resistance, and the capability of being switched off as easily as it is switched on; moreover, it can be used in both proportional and on-off configurations.

8.4.1 DC generators

When used as a power amplifier, a DC generator would have a power rating to match that of the load, and would be driven at nominally constant speed, usually by a three-phase induction motor of comparable power rating, as shown in Fig. 8.13. Typically, the power gain might be in the order of 100, whereas the voltage gain would be in the range of 1 to 10. The advantage of drawing a good current waveform from the AC mains must be offset against the capital cost of two machines to match the load, in

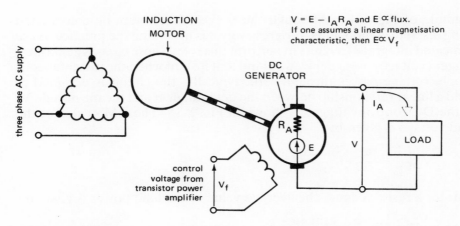

Fig 8.14 Circuit diagram and waveform for a simple single-phase Triac network

terms of power rating, together with the reliability and maintainability limitations of commutator machines.

Where the generator output drives the armature of a comparably rated DC drive motor, the system becomes a *Ward-Leonard set*. The generator field may sometimes be centre-tapped or split, to suit the requirements of the transistor power amplifier. Regenerative braking is inherent, whereby the motor can be decelerated by returning energy, via the other two machines, back into the AC mains.

8.4.2 Triac networks

A Triac network would be used where the load is a heating or lighting element. In many arrangements, the load voltage waveform would be alternating but not sinusoidal, as demonstrated in the simple single phase

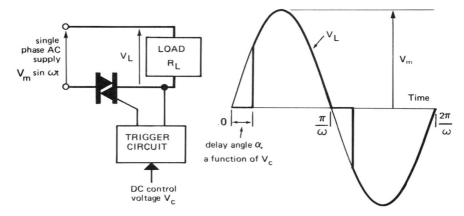

Fig 8.14 Circuit diagram and waveform for a simple single-phase Triac network

example of Fig. 8.14. Although the trigger circuit may be constructed using discrete components, progressively it is becoming the practice to use dedicated integrated circuits to perform this role. These circuits generate a trigger (voltage) pulse delayed behind that instant when the instantaneous supply voltage passes through zero: typically, the circuit is designed to yield a linear relationship between the delay angle α and the magnitude V_c of the DC control voltage, as shown in Fig. 8.15. The RMS value of the load voltage is given by

$$V_m\left(\frac{1}{2} - \frac{\alpha}{2\pi} + \frac{\sin 2\alpha}{4\pi}\right)^{1/2}$$

and, for a resistive load, the average value of the load power is given by

$$\frac{V_m^2}{R_L}\left(\frac{1}{2} - \frac{\alpha}{2\pi} + \frac{\sin 2\alpha}{4\pi}\right)$$

In Fig. 8.16 this latter function has been plotted to a base of α, and it will be seen that the relationship between power and delay angle is far from linear. With a trigger circuit having the type of characteristic shown in Fig. 8.15, the relationship between load power and control voltage V_c will be equally nonlinear.

Another type of IC trigger circuit fires the Triac in bursts of complete cycles of the AC supply, always generating the trigger pulse coincidentally with the positive-going zero-crossing of the AC supply voltage. Typically, the circuit is arranged to yield a characteristic of the form

$$\frac{\text{number of conducting cycles in a given time period}}{\text{total number of (conducting and non-conducting) cycles in the same period}} = KV_c$$

where K is a constant and V_c is the DC control voltage. Figure 8.17 shows typical load voltage waveforms, and the average value of the load RMS voltage will be proportional to the magnitude V_c of the DC control voltage.

This type of arrangement is unsuitable for loads required to react rapidly to changes in V_c, but this does not apply in most of those applications which utilise Triacs. Firing the Triac always at the instant of zero voltage crossing prevents radio frequency interference from being generated by the triggering action.

Where necessary, the trigger circuit may be isolated electrically from the gate of the Triac, using either a pulse transformer or an opto-isolator, which is a light-emitting diode and a photosensitive transistor packaged together.

The trigger circuits which have been described are also used extensively for triggering SCRs, the applications of which are covered in Sections 8.4.3, 8.4.4 and 8.4.5.

8.4.3   Converters

A suitable SCR network may be used to replace the DC generator and induction motor of the system in Fig. 8.13, and such a network, which is converting AC power into DC power, is known as a *converter*. The results from using solid state converters are an improvement in power efficiency and a distortion of the waveform of the current drawn from the AC supply.

Many alternative networks are possible, and these may be classified as follows:

- half-wave, full-wave (push-pull), or bridge
- single-phase or polyphase (including three-phase)
- fully SCR or part SCR and part power diode
- reversing or non-reversing
- regenerating (bidirectional) and non-regenerating (unidirectional).

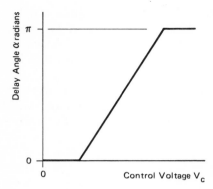

Fig 8.15   Delay angle characteristic for a typical Triac trigger circuit

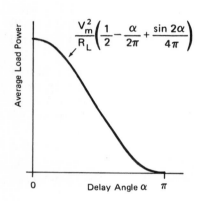

$$\frac{V_m^2}{R_L}\left(\frac{1}{2} - \frac{\alpha}{2\pi} + \frac{\sin 2\alpha}{4\pi}\right)$$

Fig 8.16   Average load power characteristic for the circuit of Fig 8.14

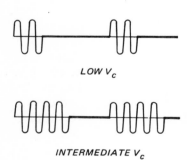

LOW V_c

INTERMEDIATE V_c

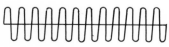

HIGH V_c

Fig 8.17   Typical load voltage waveforms for the circuit of Fig 8.14 using a zero-crossing trigger circuit

Examples of several different configurations are shown in Figures 8.18, 8.19 and 8.20. Note that the examples shown do not represent the full range of alternatives. All of the networks shown in Figs 8.18 and 8.19 may be operated in a 'bidirectional' mode: this means that, if the DC load includes a source of EMF, energy may be returned from the load to the AC supply, if the SCRs are triggered at appropriate instants in the AC voltage cycle.

All of the networks shown in Fig. 8.20 are being operated in a 'uni-directional' mode: the presence of the power diodes prevents an energy transfer from the DC load to the AC supply. Diodes D_c are known as 'commutating diodes', and they provide a short-circuit discharge path for any residual current flow through the load.

Each of the SCRs in all of the networks shown requires a suitable trigger network, and the trigger networks will require to be carefully synchronised with each other. In some cases, the gates of pairs of SCRs may simul-taneously require dual trigger pulses, when the network is switched on initially: these ensure that two SCRs are switched on, in order to effect a circuit continuity between the AC supply and the load. Usually, electrical isolation will be required between the set of trigger circuits and the set of SCRs, using pulse transformers or opto-isolators.

The circuit defined by Fig. 8.15 would be typical of those used for triggering the SCRs in these networks, in which case the characteristic relating average load power to delay angle would be nonlinear and similar to that of Fig. 8.16, although the range of delay angle may well be different from that shown.

All of the networks shown in Figs 8.18, 8.19 and 8.20 assume that the load requires a unipolar average voltage. This voltage may be made reversible, by duplicating the SCR network and its set of trigger circuits, and connecting the two networks back-to-back across the load. Figure 8.21 shows one example, using a pair of three-phase six-pulse bridges.

The waveform of the current drawn from the AC supply depends upon the 'pulse number' of the network and, as a general rule, the waveform will get closer to the ideal sinusoidal shape as this number is increased. Thus, with very high power installations, it would not be uncommon to find being used twenty-four-pulse networks, for example.

Considerations to be made when choosing an SCR converter drive include the following factors.

- What is the range of output voltage required?
- What is the range of output current required?
- Is the load centre-tapped or split?
- Does the load require a reversible signal?
- Can the output waveform be half-wave?
- Is a supply transformer to be used? If so, what will be the effect of the waveshape and the DC component of the secondary current?

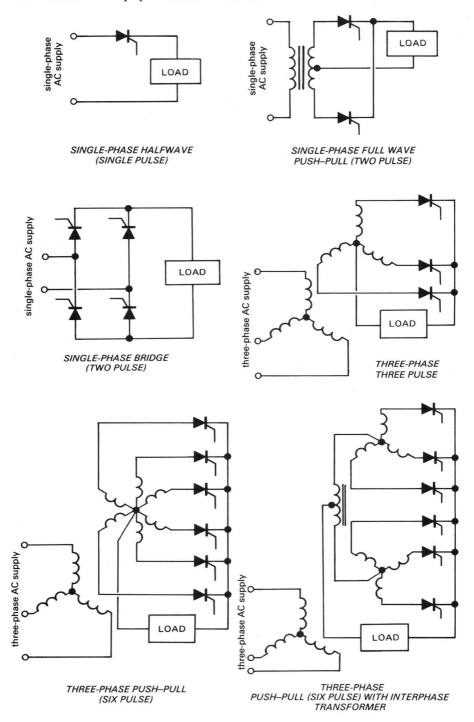

Fig 8.18 Examples of typical single-phase and three-phase SCR networks

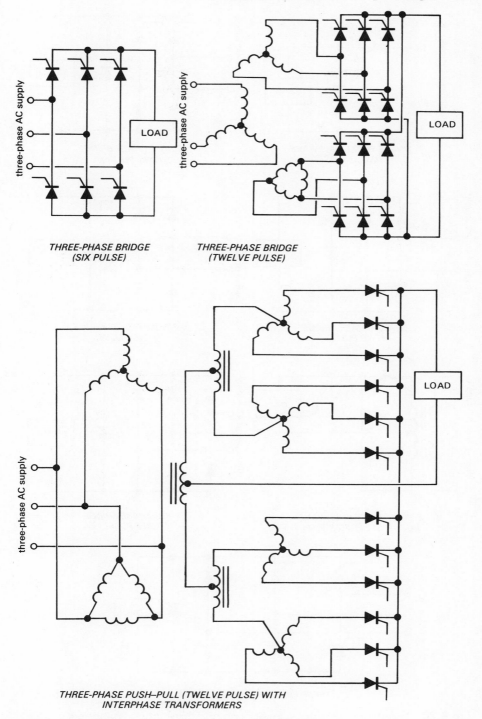

**THREE-PHASE BRIDGE
(SIX PULSE)**

**THREE-PHASE BRIDGE
(TWELVE PULSE)**

**THREE-PHASE PUSH–PULL (TWELVE PULSE) WITH
INTERPHASE TRANSFORMERS**

Fig 8.19 Further examples of three-phase SCR networks

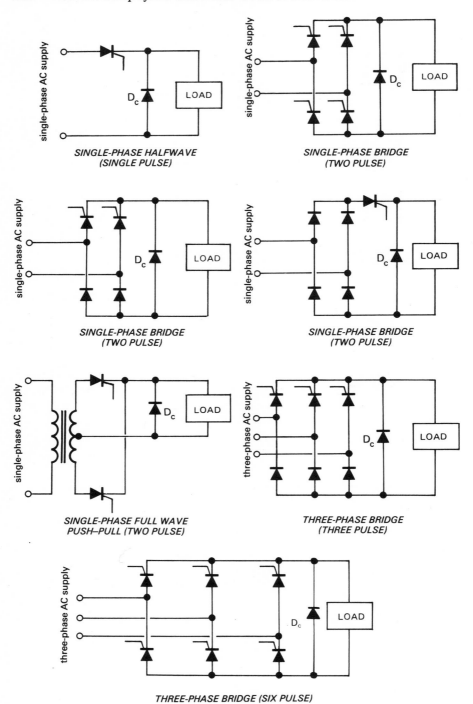

Fig 8.20   Examples of typical single-phase and three-phase networks incorporating a commutating diode to discharge residual load current

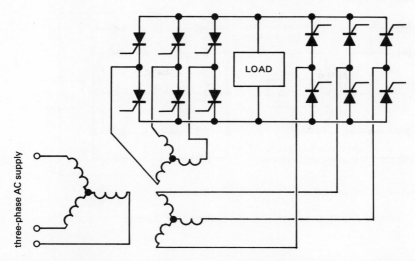

Fig 8.21   Typical three-phase six-pulse SCR bridge network for a load
requiring a reversible voltage

- Is the AC supply current waveform critical? If so, this distortion can be improved by increasing the pulse number.
- Is the load circuit to be earthed? If so, at what point?
- Are the trigger control circuits to be earthed? They may need to be isolated from the SCRs.
- Is regenerative action required? This involves arranging the triggering so as to return load energy to the AC supply.
- Is high-speed switching of large currents likely to cause interference with adjacent equipment?

8.4.4   Inverters

An *inverter* is a network for converting DC power into AC power, and therefore performs the reverse of the role of the converter. Inverters function by chopping DC voltages by various means, with the result that the alternating output waveform is synthesised and is often far from sinusoidal. Some of the factors to be considered, when selecting a suitable network for use as an inverter, include the following.

- What output voltage range is required?
- What output current range is required?
- What output frequency range is required?
- What harmonic content can be tolerated in the output voltage waveform?
- Is the load centre-tapped, with magnetic coupling between the two halves?

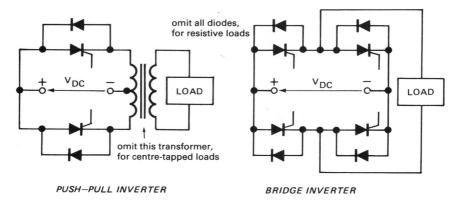

PUSH–PULL INVERTER   BRIDGE INVERTER

Fig 8.22   Basic arrangement of two alternative SCR inverter networks

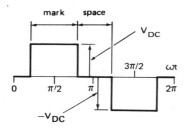

Fig 8.23   Typical load voltage waveform for the SCR bridge inverter of Fig 8.22

Most inverters use SCRs as the switching devices, in which cases special trigger circuits are required in order to switch the SCRs off as well as on. (Note that, in converters, each SCR is normally switched off by virtue of the anode-cathode voltage eventually falling to a zero value). However, some types of *gate turn-off (GTO) devices* are manufactured: these are four-layer semiconductor devices which can be switched off, as well as on, by the application of suitable triggering pulses. In addition, power MOS-FETs are most suitable as power switches, because they can be switched off and on with equal ease, so that these will be used increasingly in inverters.

Figure 8.22 shows two alternative, and basic, inverter configurations using SCRs. The triggering ('commutating') circuits have not been included in the diagrams. The SCRs are switched on and off so that full current flows through the load in one direction, in the reverse direction, or not at all, alternately: thus, the load voltage waveform will resemble Fig. 8.23, for the bridge inverter.

The RMS value of the load voltage may be controlled either by manipulating the value of V_{DC} (by generating it using a controlled converter, for example) or by varying the triggering of the SCRs, so as to manipulate the mark-space ratio. Similarly, the fundamental frequency ω of the load voltage may be varied by manipulating the frequency at which the trigger circuits generate the pulses required by the SCRs. The SCRs may be triggered in a more complex sequence, to yield different load voltage waveforms. Two representative examples are shown in Fig. 8.24, but many alternatives are possible. A significant reduction in the harmonic content of the output (which may be filtered subsequently, in any case) can result, but at the cost of increased complexity for the trigger circuits. Figure 8.25 shows one alternative arrangement for reducing the harmonic content, and this requires the establishment of stacked multiple DC supplies. Other arrangements to achieve the same type of load voltage waveform are possible.

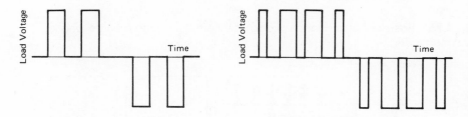

Fig 8.24   Two examples of more complex load voltage wave forms for the SCR bridge inverter network of Fig 8.22

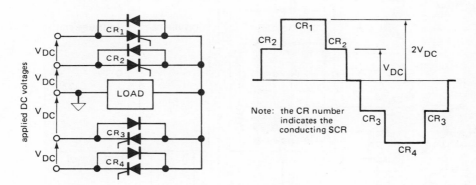

Fig 8.25   Basic circuit and load voltage waveform for a typical stacked SCR inverter

8.4.5   Frequency converters

A *frequency converter* converts AC at one voltage and frequency into AC at a different voltage and frequency. The conversion may involve an intermediate DC stage, so that the frequency converter consists of an AC–DC converter followed by a DC–AC inverter. Alternatively, the AC–AC conversion may be direct, and the network is then known as a *cycloconverter*.

Figure 8.26 shows a simple single-phase cycloconverter, which consists

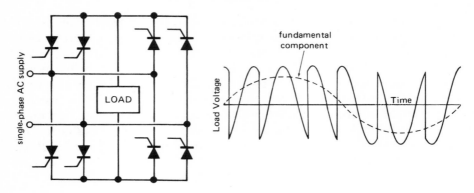

Fig 8.26   Basic circuit and load voltage waveform for a simple single-phase cycloconverter

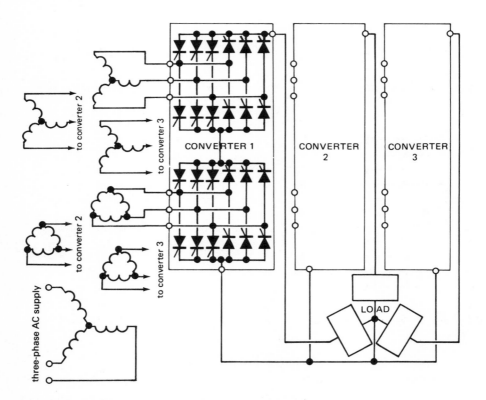

Fig 8.27   Basic circuit for a three-phase twelve-pulse cycloconverter

of two single-phase bridge converter networks, connected in opposition across the load. Figure 8.26 also shows a typical load voltage waveform together with its fundamental component. The output fundamental frequency can be seen to be a submultiple of the supply frequency, with the relationship depending upon the sequence and instants at which the SCRs are triggered on and off; this sequence will also affect the RMS value of the fundamental component. The sequence of triggering the SCRs may be organised to transfer energy in either direction.

The harmonic content of the load voltage waveform is reduced progressively as the pulse number of the cycloconverter network is increased, so that most practical networks have high pulse numbers and generally convert from three-phase to three-phase. Figure 8.27 shows a typical twelve-pulse network.

8.4.6 Magnetic amplifiers

Magnetic amplifiers were the first truly solid state power amplifiers to evolve, far in advance of semiconductor power amplifiers. They possessed many advantages, including both mechanical and electrical robustness and the ability to cope with either DC or AC input and output signals, provided that they were appropriately configured. However, because semiconductor amplifiers with far superior performance have since evolved, magnetic amplifiers will be found only in installations of long standing.

The basic building block of the magnetic amplifier is the *saturable reactor*: this physically resembles a transformer, sometimes being toroidal in form. The core is constructed from magnetic material having a hysteresis loop very square and narrow in shape, as shown in Fig. 8.28: alloys such as Hypersil, Mumetal, Permalloy and Supermalloy are suitable.

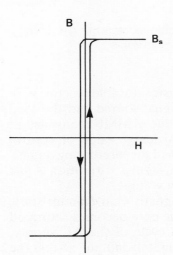

Fig 8.28 Typical hysteresis loop for the core of a saturable reactor

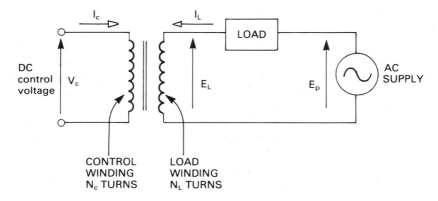

Fig 8.29   Circuit diagram for a basic saturable reactor showing signal source, power supply and load

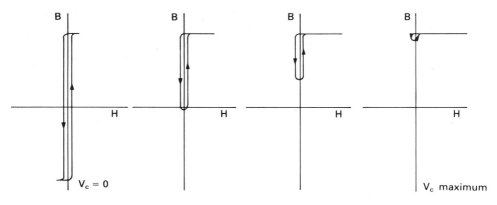

Fig 8.30   Magnetic excursion of the core of the saturable reactors of Fig 8.29 as V_e is increased positively from zero to its effective maximum value

Figure 8.29 shows the circuit diagram of the basic saturable reactor, with the control circuit being excited from a DC signal source and the load circuit excited from an AC power source. The value of MMF at any instant comprises $I_c N_c$, which is constant for any particular value of control voltage V_c, on which is algebraically superimposed the alternating quantity $i_L N_L$. E_P is of such a magnitude that, when I_c is zero, saturation is just avoided: E_P is the RMS value of the AC supply voltage.

As $I_c N_c$ is raised, the working point on the magnetisation charcteristic is raised from the origin, as shown in Fig. 8.30: the core becomes saturated for a progressively greater proportion of each cycle of the AC supply. When the core is unsaturated, dB/dH is approximately infinite, so that the impedance of the load winding will be very high and i_L approximately zero. Thus, $e_L \cong e_P$, in this region.

When the core is saturated, dB/dH is approximately zero, so that the impedance of the load winding will be very low and e_L approximately zero. Thus, $i_L R_L \cong e_P$, in this region.

In a sense, the behaviour is similar to that of an SCR, because the control voltage is causing the controlling device to switch between very high and very low impedance states. The reactor acts as a form of switch: when its core is unsaturated, the AC voltage appears across the load winding; when its core is saturated, the AC voltage appears across the load. The control voltage determines the relative proportion of each AC supply cycle in which the load circuit is in each state.

The description of the operation of the reactor is over-simplified, so far, because it ignores the fact that the AC in the load circuit will, by transformer action, be reflected back into the control circuit: this effect needs to be suppressed, particularly since, in practice, N_c is very much greater than N_L.

The technique usually adopted to minimise the reflected AC is to use duplicated saturable reactors and to interconnect them in such a way as to maximise possible cancellation of this effect: reactors paired in this way then are called *magnetic amplifiers*. In most applications, magnetic amplifiers are used to control DC loads, necessitating the full-wave rectification and smoothing of the AC current flowing in the load windings. Figure 8.31 shows two of many possible alternative circuit configurations.

Figure 8.32 shows the nature of the transfer characteristic for amplifiers of the type depicted in Fig. 8.31. Clearly, these arrangements are inadequate if the load requires a reversible DC current: such a requirement can be satisfied by using two balanced networks like one of those shown in

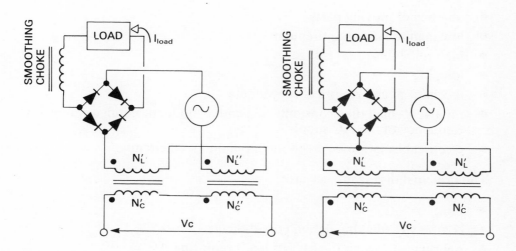

Fig 8.31   Two of many alternative configurations for interconnecting two saturable reactors to form a magnetic amplifier having a DC load

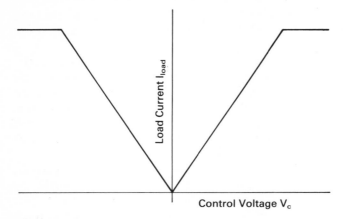

Fig 8.32 Idealised transfer characteristic (static characteristic) for magnetic amplifier configurations such as those depicted in Fig 8.31

Fig. 8.31. Figure 8.33 shows an example of such a bipolar configuration and, again, this is just one of many possible alternative arrangements.

Section 8.5 explains various reasons for wishing to complete negative feedback loops around stages of amplification. In the case of magnetic amplifiers, one relatively simple way to implement this is to pass the load current back through extra control windings added to the saturable reactors. It will be seen by this time that a magnetic amplifier driving a load requiring a controlled reversible DC and incorporating negative feedback will be a pretty complicated arrangement.

In summary, the principal advantages with magnetic amplifiers are:

- absence of moving parts
- low maintenance requirement
- high reliability
- long life
- mechanical and electrical ruggedness
- relatively small time constants, depending upon the configuration and frequency of the AC supply
- electrical isolation between the input and load circuits.

The principal disadvantages are:

- bulk
- complexity, making design very difficult
- poor waveform, in the case of loads requiring AC
- relatively low power efficiency in comparison to equivalent semiconductor amplifiers.

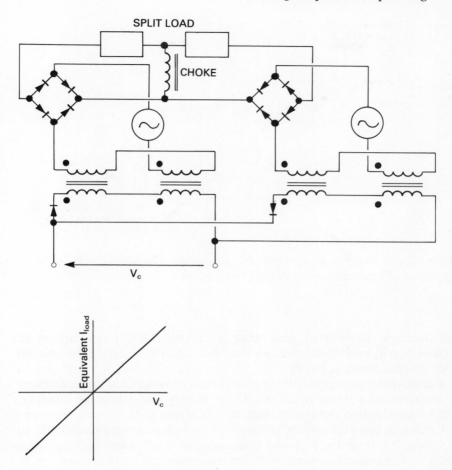

Fig 8.33   An example of a magnetic amplifier configuration driving a load requiring bipolar DC

8.4.7   Rotating amplifiers

Special DC generators were developed, in the past, to function specifically as power amplifiers; however, because semiconductor power amplifiers with far superior performance have since evolved, these machines will only be found in installations of long standing. Such special machines fall within a category given the alternative generic names of *rotating amplifiers* and *cross-field generators*, and have been allotted a variety of trade names, such as Metadyne, Amplidyne, Rototrol, Magnavolt, etc.

Figure 8.34 demonstrates the principle of the cross-field generator, which is equivalent to two separately excited conventional DC generators connected in cascade. In a DC generator, the armature current is very much greater than the field current. Suppose that the current ratio is 100:1.

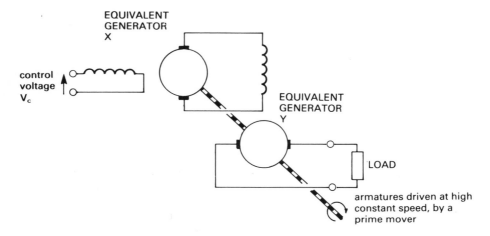

Fig 8.34   Representation of a cross-field generator as two equivalent conventional DC generators connected in cascade

When the field current of generator X is 1 mA, the field current of generator Y will be 100 mA and its load current will be 10 A: the overall current amplification is 10 000.

In a cross-field generator, the two equivalent machines are combined into the frame of a single machine. There is only one armature winding, within which the two equivalent armature current patterns are superimposed: this is achieved by using two sets of brushes, mounted electrically in quadrature, running on a single commutator. The poles and the field winding of equivalent generator Y are not essential for the machine to function, because the armature reaction flux of equivalent generator X could be used to perform the function of the main flux of Y. This flux is given a significant magnitude, typically by short-circuiting the brushes of X upon themselves, to maximise the magnitude of the armature current pattern establishing this flux. The arrangement is represented symbolically in Fig. 8.35. When load is connected to the machine, the armature current pattern associated with equivalent generator Y establishes its own armature reaction flux, and this will act along an axis aligned with that of the main flux of X and in opposition to it. The function of the compensating winding shown in Fig. 8.35 is to enable the load current to develop additional MMF to cancel partially or completely the armature reaction MMF which it has established. The nett flux acting along the control axis is then either mainly or completely due to the MMF associated with the control field current. It can be shown that, with 0% compensation ($N_{com} = 0$)

$$I_L = V_c \cdot \frac{N_c}{N_a R_c}$$

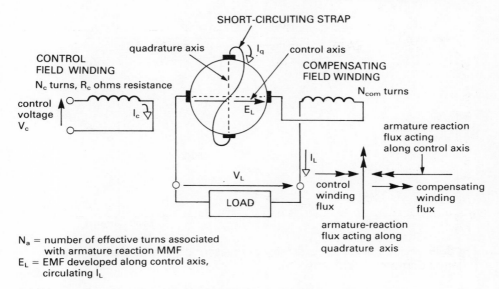

Fig 8.35 Symbolic representation of a cross-field generator

so that the machine behaves approximately as a constant current source, for a given value of control voltage. Additionally, with 100% compensation ($N_{com} = N_a$)

$$E_L = V_c \cdot \frac{KN_c}{R_c}$$

so that the machine now behaves approximately as a constant voltage source, for a given value of control voltage. K is a constant of proportionality, the value of which is maximised by running the machine at high speed: a figure of several thousand rev/min would be typical, so that the prime mover would usually need to be a DC motor.

Figure 8.36 shows a typical set of load characteristic curves for a cross-field generator, for varying degrees of compensation and a specific value of control voltage.

Figure 8.37 shows a more complete circuit diagram for a typical cross-field generator. The additional windings on the control axis permit the addition and subtraction of input signals whilst affording electrical isolation for the sources of those signals: see Section 12.2.8. The Ampliator winding permits high levels of quadrature axis MMF whilst lowering the level of quadrature armature current I_q.

The principal advantages with a cross-field generator are the high level of power gain which can be achieved, time constant values which are very small for an electrical machine, and electrical isolation between all input and output circuits. The principal disadvantages are those associated with the maintenance of a commutator machine, particularly one running at high speed and in which the development of significant levels of armature

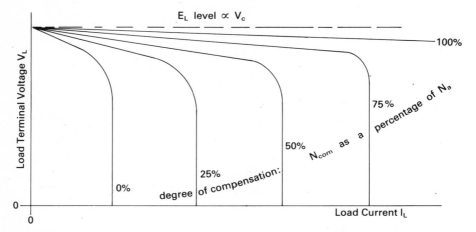

Fig 8.36 Family of load characteristic curves for a cross-field generator for different degrees of compensation

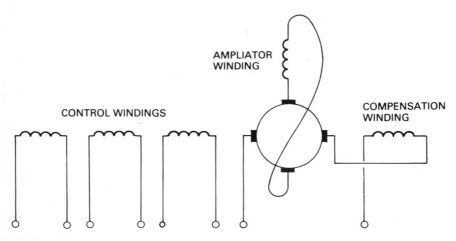

Fig 8.37 More complete circuit diagram for a typical cross-field generator

reaction is intentional (resulting in much brush arcing); in addition, of course, the use of a cross-field generator also necessitates the installation of a prime mover having a comparable power rating. However, one could mechanically couple several cross-field generators to a single prime mover of suitable size, where appropriate.

8.5   The use of minor negative feedback loops

Often, it is found that (minor) negative feedback loops have been incorporated around one or more elements in the forward path of a control system. There are several possible reasons why this should be the case, and these will be explained.

8.5.1 Negative feedback to linearise a nonlinear static characteristic

Referring to Fig. 8.38, suppose that the element with a gain B has significant nonlinearities in its static characteristic. It is possible to produce a closed loop having a much more linear static characteristic, by adding gain A to the forward path and a linear network H to the feedback path, as shown in the figure. The gradient of the nonlinear characteristic B will, at the operating point, be given by $\partial V_o / \partial V_a$.

Figure 8.39 is a small-signal block diagram for the loop, showing increments in all of the voltages.

Now

$$\delta V_e = \frac{1}{A} \cdot \delta V_a = \delta V_i - H \cdot \delta V_o \quad \text{and} \quad \delta V_o = \frac{\partial V_o}{\partial V_a} \cdot \delta V_a$$

so that

$$\frac{1}{A} \cdot \frac{1}{\dfrac{\partial V_o}{\partial V_a}} \cdot \delta V_o = \delta V_i - H . \delta V_o$$

whence

$$\delta V_o = A \cdot \frac{\partial V_o}{\partial V_a} \cdot \delta V_i - AH \cdot \frac{\partial V_o}{\partial V_a} \cdot \delta V_o$$

Finally

$$\frac{\partial V_o}{\partial V_i} = \lim_{\delta V_i \to 0} \left[\frac{\delta V_o}{\delta V_i} \right] = \lim_{\delta V_i \to 0} \left[\frac{A \cdot \dfrac{\partial V_o}{\partial V_a}}{1 + AH \cdot \dfrac{\partial V_o}{\partial V_a}} \right]$$

$$= \lim_{\delta V_i \to 0} \left[\frac{A}{\dfrac{1}{\dfrac{\partial V_o}{\partial V_a}} + AH} \right]$$

If

$$AH \gg \left(\frac{\partial V_o}{\partial V_a} \right)^{-1}$$

then

$$\frac{\partial V_o}{\partial V_i} \cong \frac{A}{AH} = \frac{1}{H}$$

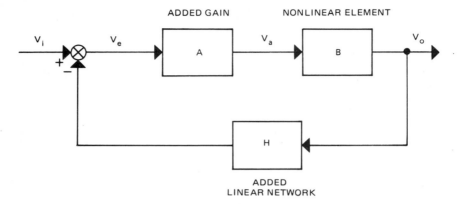

Fig 8.38 Block diagram showing a negative feedback loop created to disguise the nonlinear characteristic of B, by the addition of forward path gain A and linear feedback path network H

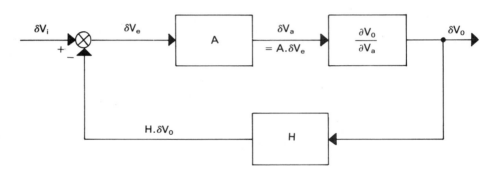

Fig 8.39 Small-signal block diagram for the loop of Fig 8.38

However, when $\partial V_o/\partial V_a = 0$, corresponding to the nonlinear element being operated in a region of saturation, $\partial V_o/\partial V_i = 0$ whatever the values of A and H.

The insensitivity of the closed loop incremental gain to changes in the incremental gain of the nonlinear element improves as the product AH is progressively increased: for any particular operating point on the B characteristic, other than one within a region of saturation, the linearity of the closed loop static characteristic will be improved by a factor of approximately ten for every ten-fold increase in AH. Since the addition of the elements A and H has changed the gain from the original (nonlinear) value of B to one of approximately 1/H, then, if the original order of gain value is still required, appropriate amplification (or attenuation) may be added, ahead of the closed loop.

Negative feedback will have a linearising effect on all nonlinear properties except saturation. If physical considerations limit the excursion of the output V_o of element B, then no amount of feedback will change this limit: $\partial V_o / \partial V_i = 0$ whenever $\partial V_o / \partial V_a = 0$, irrespective of the magnitudes of A and H. However, the amount of feedback *will* determine the value of input Signal V_i needed to drive the output V_o to its limit.

The shape of the closed loop static characteristic can be predicted as follows. Suppose that Fig. 8.40 represents the static characteristic of element B. This can be plotted to a base of V_a/A, for the chosen value of A, as shown in Fig. 8.41. Superimposed on this is a plot of V_o vs HV_o, for the chosen value of H. Now $V_i = (HV_o + V_e) = (HV_o + V_a/A)$ so that, for any arbitrary value of V_o, the corresponding value of V_i may be determined by adding the two horizontal ordinates x_1 and x_2, as shown in

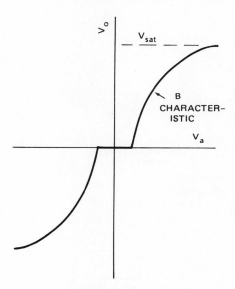

Fig 8.40   Representative static characteristic of element B of Fig 8.38

Fig. 8.41. For the given example, the final closed loop static characteristic would resemble Fig. 8.42. Note that the output V_o still saturates at the original level V_{sat}.

8.5.2   Negative feedback to enhance the speed of response of an element

Suppose that a linear element has a DC gain K and a simple lag time constant of T seconds, as shown in Fig. 8.43. Then its unit step response will be given by

$$V_o(t) = K(1 - e^{-t/T})$$

If gain A and linear feedback H (both assumed to be frequency indepen-

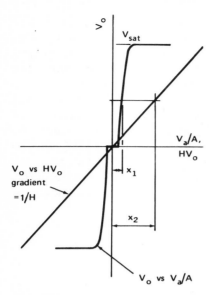

Fig 8.41   Construction for developing the closed loop static characteristic for the loop in Fig 8.38

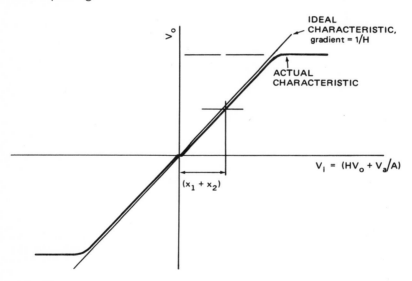

Fig 8.42   Closed loop static characteristic for the system defined by Fig 8.38 and 8.40

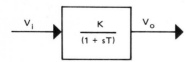

Fig 8.43   Block representing gain plus a simple lag

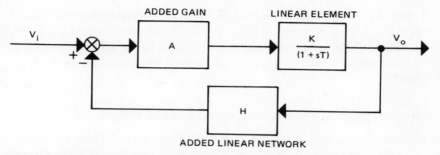

Fig 8.44   Feedback loop closed around the block in Fig 8.43

dent) are added, as shown in Fig. 8.44, this closed loop configuration will then have a unit step response given by

$$V_o(t) = \frac{AK}{(1 + AHK)}\,(1 - e^{-t(1 + AHK)/T})$$

Thus, the DC gain and the lag time constant have effectively been reduced by the factor $A/(1 + AHK)$ and $1/(1 + AHK)$ respectively, so that, if $AKH \gg 1$, a considerable reduction in time constant will result. The level of DC gain may be restored by adding amplification of value $(1 + AHK/A)$ externally to the loop: if $A \gg 1$, the added gain will need to have a value of HK, approximately.

8.5.3   Negative feedback to change a voltage source into a current source

Consider the configuration of Fig. 8.45, in which the output of amplifier A behaves as a voltage source and dropping resistor R has been added in series with the load.

Now $V_o = A(V_i - I_L R) = I_L(R + R_L)$

$$AV_i = I_L(R + R_L + AR)$$

$$\frac{I_L}{V_i} = \frac{A}{(1 + A)R + R_L}$$

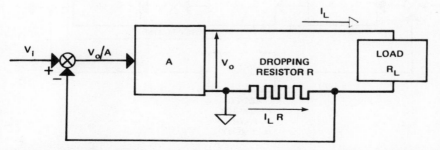

Fig 8.45   Technique for converting a voltage source into a voltage-controlled current source

If $A \gg 1$ and $AR \gg R_L$ then $I_L/V_i \cong 1/R$. The relationship between load current and loop input voltage is therefore a linear one, since $I_L \cong V_i/R$, and is virtually independent of possible variations in R_L and A. The loop therefore behaves as a voltage-controlled current source, as far as the load is concerned.

8.5.4   Negative feedback to improve signal-to-noise ratio

Where parasitic noise occurs within a control element, its effect within the system can often be reduced by adding a high-gain negative feedback loop around that element. Frequently, one or more of the added elements would be frequency dependent, so that it introduces filtering action at the noise frequencies. Analysis of Fig. 8.46 yields the following relationships:

Signal gain without the added elements $\dfrac{C}{R}(j\omega) = G_2 G_3(j\omega)$

Noise gain without the added elements $\dfrac{C}{N}(j\omega) = G_3(j\omega)$

Signal gain with the added elements $\quad \dfrac{C}{R}(j\omega) = \dfrac{G_1 G_2 G_3(j\omega)}{[1 + G_1 G_2 G_3 G_4(j\omega)]}$

Noise gain with the added elements $\quad \dfrac{C}{N}(j\omega) = \dfrac{G_3(j\omega)}{[1 + G_1 G_2 G_3 G_4(j\omega)]}$

The effect of the added elements has been to reduce the spectrum of the noise component of the output by the factor $[1 + G_1 G_2 G_3 G_4(j\omega)]$, whereas the spectrum of the signal component of the output has only been

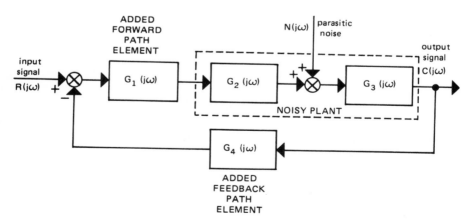

Fig 8.46   Application of negative feedback to improve signal-to-noise ratio within a plant process

modified by the factor $G_1(j\omega)/[1 + G_1 G_2 G_3 G_4(j\omega)]$. The relationship between the two spectra will depend largely upon the choice of the $G_1(j\omega)$ frequency response.

8.6 Servomotors

Motors which are called specifically *servomotors* have been especially developed for use in position servosystems. Because of the nature of the market, they have been manufactured only in small power ratings so that, where higher power levels are required, it is then necessary to use conventional motors. The most common servomotors are the cage AC servomotor, the DC servomotor and the *stepper motor*; however, there are some less common alternative types of small motor used in servo-systems.

8.6.1 AC servomotor

This is usually a two-phase cage rotor type of induction motor and it is operated as indicated in Fig. 8.47. The control winding is supplied with an AC voltage of varying magnitude V_c, delivered by an AC servoamplifier. The reference winding is supplied with a quadrature AC voltage of fixed magnitude V_{ref}, delivered by an AC power source . Typically, the frequency is 50, 60 or 400 Hz and synchronous speed will be proportional to the frequency chosen. The voltage levels will be such that $V_c \leqslant V_{ref}$.

Figure 8.48 is a phasor diagram of the voltages on which V_{ref} has been resolved into two components: one equal in value to $| V_c |$ and which, when taken in combination with V_c, will result in polyphase action; the other is equal to $(V_{ref} - | V_c |)$ and will result in single-phase action. When single phasing ($V_c = 0$), the motor flux Φ_{ref} is a pulsating one, aligned along an axis stationary in space. Such a flux may be resolved into two components,

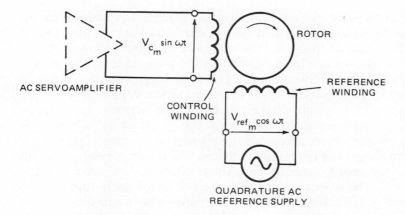

Fig 8.47 Usual electrical connections for a two-phase AC servomotor

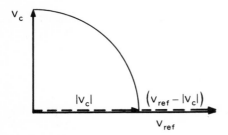

Fig 8.48   Phasor diagram for the AC voltages applied to two-phase AC servomotors

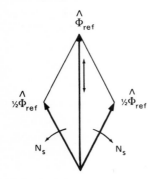

Fig 8.49   Resolution of the flux vectors for the single phasing case

constant in magnitude and equal to half the peak value of Φ_{ref} rotating in opposite directions at synchronous speed, as indicated in Fig. 8.49. Each component of flux is responsible for a polyphase type of action and these two actions, which will act in opposition to one another, can be combined graphically, as shown in Fig. 8.50 for a conventional type of induction motor.

In the AC servomotor, the rotor circuit is designed to have an abnormally high resistance, by using high resistivity material for the cage bars and/or end rings. The effect of this is to 'stretch' each component torque curve horizontally, as shown in Fig. 8.51. The single phasing torque is now a braking torque, because the direction (sign) of the torque is always in opposition to the direction of the motion (sign of the velocity).

When $V_{ref} > V_c > 0$, single phase and polyphase actions occur simultaneously, in proportions depending upon the value of V_c: when V_c is small, single-phase action predominates; when V_c is large, polyphase action predominates. Using two phase symmetrical components, one can show that the pure polyphase and pure single phase actions are related in the ratio of 4:1. The overall result of these relationships is summarised graphically in Fig. 8.52. Manufacturers normally publish one quadrant of this graph, for torque and speed both positive. The set of characteristics which results may be idealised by a set of parallel straight lines, equally

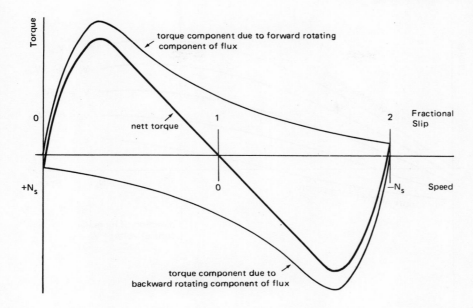

Fig 8.50 Construction for producing the torque vs speed characteristic of a conventional single-phase induction motor

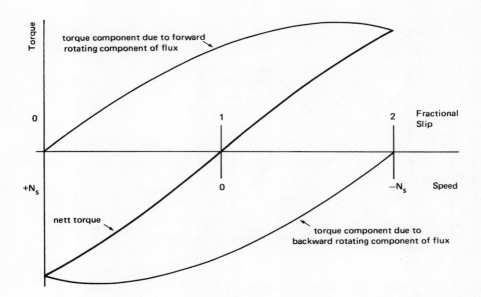

Fig 8.51 Construction for producing the torque vs speed characteristic of an AC servomotor when single-phasing

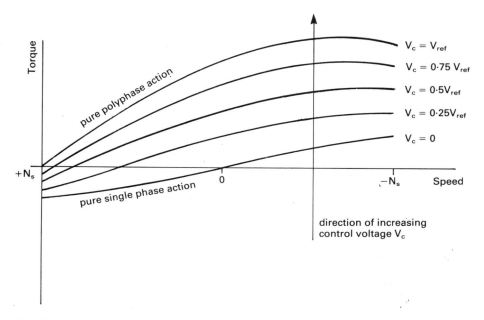

Fig 8.52   Family of torque vs speed characteristic for a two-phase AC servomotor

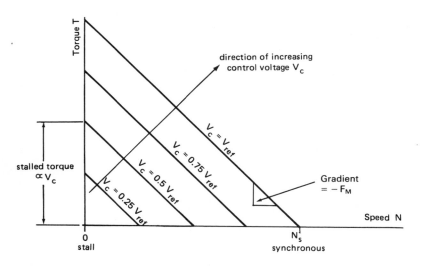

Fig 8.53   Idealised and linearised representation of torque vs speed characteristics for a two-phase AC servomotor, for torque and speed both positive

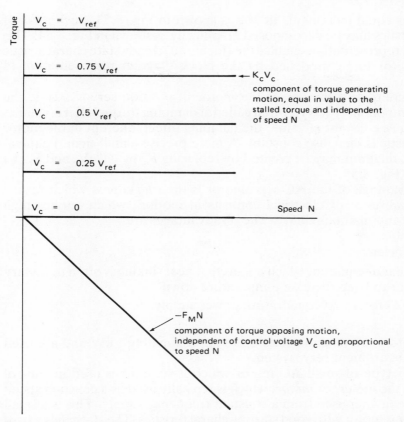

Fig 8.54   Decomposition of the idealised AC servomotor characteristics of Fig 8.53

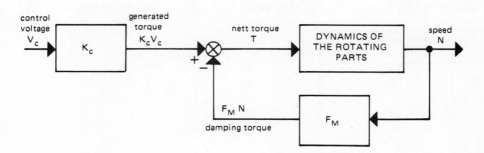

Fig 8.55   Block diagram representation of an AC servomotor corresponding to the torque vs speed characteristics of Fig 8.53 and 8.54

spaced for equal increments in V_c, as shown in Fig. 8.53. Such a set of characteristics may be decomposed graphically as shown in Fig. 8.54. This linearised representation enables the (idealised) steady state characteristic of the motor to be modelled by the block diagram representation of Fig. 8.55.

In the context of the use of the motor in position servosystems, the torque component $(F_M N)$ is often called a 'damping torque'. If it is valid to regard F_M as constant in value, the damping effect inherent in the motor characteristic is then also constant. A more precise (small signal) equivalent block diagram may be created by replacing K_c by $\partial T/\partial V_c$ and F_M by $\partial T/\partial N$, in Fig. 8.55.

The advantage of the AC servomotor is the ruggedness which results from the absence of electrical contacts in motion, which implies high reliability and maintainability. The disadvantages are:

- low efficiency
- high heat dissipation, which means that heat sinking is often necessary in order to keep the case temperature down
- the need for an AC quadrature power supply.

These motors are made in sizes up to approximately 1 kW and are used mainly in instrument servosystems.

Another type of small AC motor which is sometimes used in control systems is the *hysteresis motor*, which is typically used as a constant speed source, when energised from a constant frequency supply. This is a small synchronous motor with good starting characteristics. The stator has a two-phase winding which is excited conventionally. The rotor is a cylinder of magnetic material of the type normally used for permanent magnets. The stator field rotates at synchronous speed, due to pure polyphase action, and this induces a similar field in the rotor, when it is stationary. However, because of hysteresis effects, the rotor field lags behind the stator field and

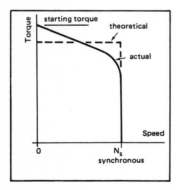

Fig 8.56 Theoretical and actual torque vs speed characteristic for a Hysteresis motor

this lag causes a torque to be generated: this torque accelerates the rotor and load in the same direction as the field rotation. Theoretically, this situation is sustained at all speeds up to synchronous speed, which is the steady-state speed of the motor. In practice, the starting torque is somewhat enhanced by induction motor action. Figure 8.56 shows a typical torque vs speed characteristic.

8.6.2   DC servomotors

These motors are constructed like a conventional DC motor, but in miniature and with a high length/diameter ratio for the armature, in order to minimise the polar moment of inertia. The field may be wound, in which case it may be split or centre-tapped. Alternatively, the field system may be established by permanent magnets (often ferrites), in which case the motor is often called a *PM motor* and only armature control is possible. The armature and commutator may be a heavy-duty double-sided printed circuit, and such a motor is often called a *PC motor*: the low armature mass offsets the low length/diameter ratio, to keep the polar moment of inertia low in value.

When operated in a servosystem, the motor may be either field controlled or armature controlled, and the armature may be fed from either a voltage source or a current source. Each combination yields a different family of torque vs speed characteristics. With split field DC servomotors, which have wound field systems, the field coils may be interconnected, internally, in such a way that coils connected in one electronic circuit are associated with a magnetic circuit different from that with which the remaining coils are associated. This is shown symbolically in Fig. 8.57.

The left-hand part of Fig. 8.58 shows how the magnetic fluxes in the two magnetic circuits vary as the control voltage V_c is varied. This diagram assumes that the servoamplifier output stage is biased for Class B operation, so that the currents in the two field circuits are both zero when V_c is zero. This diagram also shows the nominal hysteresis loop for each magnetic circuit, which would result from full bipolar excursion of the field current.

The right-hand side of Fig. 8.58 shows the variation in equivalent composite magnetic flux, obtained by adding together the individual hysteresis loops for Φ_{12} and Φ_{43}. It will be seen that the nett effect of this split field arrangement, when compared with the effect which would occur with a simple single field arrangement (as represented by the nominal hysteresis loop), is as follows:

- the nett flux is zero when the control voltage is zero, so that the generated torque is also zero
- there is no hysteresis (and therefore remanent magnetism) present at zero control voltage
- the hysteresis present at other values of V_c will be less than that in the single field case.

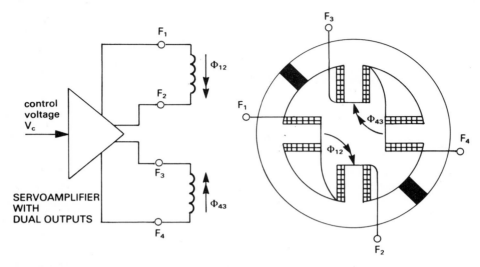

Fig 8.57   Electrical and physical orientation of the field windings of a split field DC servomotor

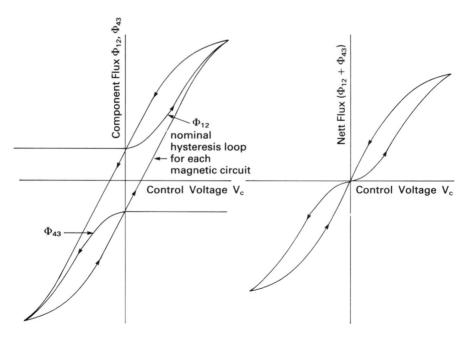

Fig 8.58   Component values of excitation and nett excitation for a split field motor

It can readily be shown that the improvement resulting from the split field arrangement will not be as great if the servoamplifier is biased differently from Class B operation, for which the two field currents are zero simultaneously when the control voltage is zero.

Field control of the motor gives rise to the following comments:

- it is possible only with wound field motors
- the power rating of the field control amplifier need match only the (low) power requirements of the field
- the field time constant will be relevant to the dynamic response of the motor
- the magnetisation characteristic will be relevant to the linearity of the behaviour.

Armature control gives rise to the following comments:

- the field system can be wound or permanent magnet
- the power rating of the armature control amplifier must match the (relatively high) power requirements of the armature
- the field time constant and magnetisation characteristic are not relevant to the system dynamic performance.

When the armature is supplied from a voltage source, it is often necessary to incorporate a current limit facility within this source. This is often because damagingly high armature currents may otherwise circulate whenever the back-EMF is significantly different in value from the terminal voltage. The current limit circuit may incorporate a time delay, enabling high transient armature currents to circulate for short time durations.

In the analysis which follows, Φ represents magnetic flux, T represents torque and N represents shaft speed; the Ks are constants of proportionality and a linear magnetisation characteristic is assumed.

Field-control, constant armature voltage supply

Figure 8.59 represents this type of configuration.

$$\Phi = K_\Phi I_f = K_\Phi \frac{V_f}{R_f}$$

$$T = K_T \Phi I_A$$

$$= K_T K_\Phi \frac{V_f}{R_f} \cdot \frac{(V - E)}{R_A}$$

$$E = \frac{N\Phi}{K_N} = \frac{N K_\Phi V_f}{K_N R_f}$$

$$T = \frac{K_T K_\Phi V_f}{R_f R_A}\left(V - N\frac{K_\Phi V_f}{K_N R_f}\right) = K_1 V_f - K_2 N V_f^2$$

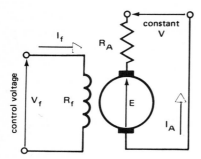

Fig 8.59   Circuit configuration of field control of a DC servomotor with a constant armature voltage supply

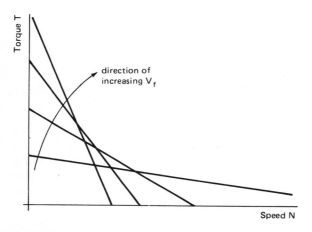

Fig 8.60   Family of torque vs speed characteristics for a field-controlled DC servomotor with constant armature voltage

Thus, the stalled torque is proportional to V_f and the damping torque is proportional to V_f^2 (so that the damping effect is variable with control signal level), as well as speed N. The characteristic is represented in Fig. 8.60.

Field control, constant armature current supply
   Figure 8.61 represents this type of configuration.

$$\Phi = K_\Phi I_f = K_\Phi \frac{V_f}{R_f}, \qquad T = K_T \Phi I_A = K_T K_\Phi \frac{V_f}{R_f} I_A = K_3 V_f$$

Thus, the torque is proportional to V_f and is always independent of speed N. The damping effect is zero. The resulting characteristic is shown in Fig. 8.62.

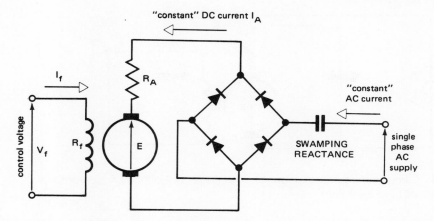

Fig 8.61 Circuit configuration of field control of a DC servomotor with a nominally constant armature current supply

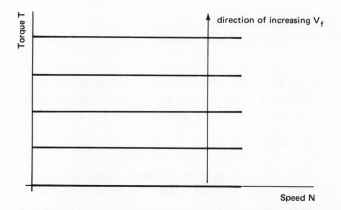

Fig 8.62 Family of torque vs speed characteristics for a field-controlled DC servomotor with constant armature current

Armature voltage control, constant excitation

Figure 8.63 represents this type of configuration.

$$E = \frac{N\Phi}{K_N}$$

$$T = K_T\Phi I_A = K_T\Phi\frac{(V - E)}{R_A} = K_T\Phi\left(\frac{V}{R_A} - \frac{N\Phi}{K_N R_A}\right)$$

$$= K_4 V - K_5 N$$

The stalled torque is now proportional to V and the damping torque is proportional to speed N alone, so that the damping effect is constant. The torque vs speed characteristics will now resemble Fig. 8.53, but the

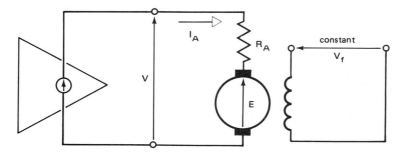

Fig 8.63   Circuit configuration of armature voltage control of a DC servomotor with constant field excitation

gradient will be $-K_5$ and the control voltage will be the armature voltage V.

Armature current control, constant excitation

Figure 8.64 represents this type of configuration.

$$T = K_T \Phi I_A = K_6 I_A$$

The torque is proportional to I_A and is always independent of speed N. The torque vs speed characteristics will resemble Fig. 8.62, except that the control signal will now be I_A. The damping effect is zero.

Further comments on DC servomotors

The advantages with DC servomotors are that they employ DC control signals and that a range of alternative torque vs speed characteristics is available, depending upon the mode of control. The disadvantages accrue from the use of a commutator: reliability and maintainability will be low and there may be problems arising from radio-frequency interference generated by brush arcing.

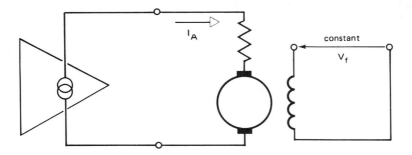

Fig 8.64   Circuit configuration of armature current control of a DC servomotor with constant field excitation

In high-power servosystems, the servomotors (AC or DC) may be used as input drives to hydraulic servovalves. In many cases, the angular travel may be extremely small (the motor may be driving against restraining springs), so that the primary function of the motor is to develop torque rather than achieve significant motion: the motor in this context is generally called a *torque motor*. Simpler types of torque motor also are constructed, using DC excited coils and ferrous magnetic circuits incorporating permanent magnets: these will be covered in Section 8.6.4.

8.6.3 Stepper motors

Stepper motors have been used principally for drives in computer peripherals: hard and soft disc head positioners, paper tape drives, incremental magnetic tape drives, printer drives, card reader drives, etc. Mostly, they have been small in physical size but developing a high torque/inertia ratio and a fast response. They have also been used as torque motors for hydraulic servovalves in high-power servosystems. More recently, high-power stepper motors have been developed for industrial applications, such as machine tools, industrial robots, automatic draughting machines, robot cameras, etc.

A stepper motor has a stator containing a number of poles carrying field windings, together with a rotor which, in most cases, is a permanent magnet or which is slotted and made from soft iron. By energising the stator windings in sequence, the axis of the magnetic field is stepped around and the rotor axis tracks this motion.

The drive circuits normally accept a train of pulses, each pulse representing one quantum step in shaft angle, together with a signal indicating the direction of motion required. The circuits use this information to establish a pattern of voltages, which is then applied appropriately to the stator windings.

These motors are unique in the sense that the control signals determine the rotor position, as opposed to the rotor speed, so that the motors can be used in position servosystems without the necessity for position feedback transducers. The motors are relatively inefficient and often require extensive heat sinking to remove heat.

The advantages of stepper motors, when contrasted with DC servomotors, are as follows:

- steppers do not require linear power amplifiers
- feedback transducers are optional
- steppers can respond directly to digital control data
- steppers exhibit extremely short starting and stopping times
- steppers have a very wide speed range.

The disadvantages of the stepper motor are as follows:

- it is the most difficult of all motor types to analyse and specify

- it has low power efficiency, associated with high internal heat generation

- it displays a tendency to oscillate about the equilibrium point.

Figure 8.65 shows the principles of construction of three different types of stepper motor.

With the *permanent magnet* type of stepper motor, the rotor is a permanent magnet. The rotor aligns its magnetic axis with the axis of the magnetic field generated by the DC current being passed through the coils on the stator poles. When the coils are energised in sequence, the rotor steps around to track the motion of the stator field. The rotor is rather

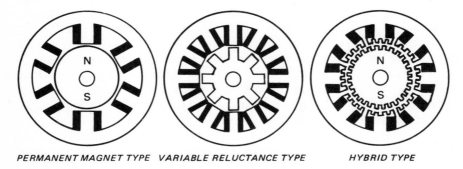

PERMANENT MAGNET TYPE   VARIABLE RELUCTANCE TYPE   HYBRID TYPE

Fig 8.65   Symbolic representation of cross-sections through the three principal types of stepper motor

large, in order that an adequate field can be developed: this gives rise to a relatively high polar moment of inertia and a low maximum stepping rate (typically 300 to 400 steps/second). Because of the stator construction, the angular step size will be large: typically 30° or 60°. Damping is inherent, due to generator action which induces EMFs in the stator coils whenever the rotor tends to oscillate. This type of stepper exhibits a high holding torque even without sustained stator energisation, and the stepping torque is proportional to the stator current.

With the *variable reluctance* type of stepper motor, the rotor is made from unmagnetised soft iron and is cylindrical but with slots. Changes in rotor position vary the reluctance of the flux path between stator poles having opposite polarities. The rotor always moves to a radial alignment of minimum reluctance, which depends upon which particular stator windings happen to be energised. The rotor can be small and light, and the resulting low polar moment of inertia yields a high maximum stepping rate (typically 700 to 800 steps/second) and fast starting and stopping. Because of the rotor construction, the angular step size will be smaller than with the permanent magnet type: typically 15°. The damping effect is negligible, due to the absence of any generator action. The variable reluctance type exhibits no holding torque when the stator is de-energised, so that a minimum stator energisation would normally be required at standstill. The stepping torque is proportional to the square of the stator current.

The *hybrid* type of stepper motor has a slotted soft iron rotor but includes a permanent magnet in its magnetic circuit: usually the magnet forms part of the rotor construction. Hybrid stepper motors generally have a high stepping torque, small step size (typically 0.5° to 15°), high polar moment of inertia, and a relatively low maximum stepping rate (typically 150 to 250 steps/second).

Some speed controllers for stepper motors are described in Section 11.4.3, whilst further discussion on the use of steppers for position control and speed control is contained in Sections 13.4.3 and 13.5.3, respectively.

8.6.4 Torque motors and force motors

A torque motor can be created by employing a servomotor to develop torque, whilst restraining it from moving by more than a few degrees of total displacement. The motor therefore always operates close to or in the stalled condition, and must be rated accordingly.

Such an application is wasteful for an expensive servomotor, so that it is normal to use a simpler and cheaper construction for most torque motors. A force motor operates using similar principles to those of simple torque motors, except that its motion is rectilinear rather than rotary.

The electromagnetic torque motor has a magnetic circuit created by permanent magnets and pole pieces of high magnetic permeability material. An armature, of similar magnetic permeability material, is placed between the pole pieces. Around this armature a coil is placed in order to manipulate the value of the magnetic flux.

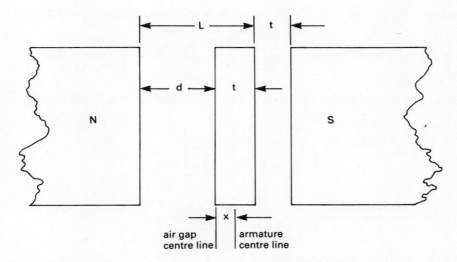

Fig 8.66 Schematic arrangement of a torque motor pole pieces and armature

To explain the principle of operation, consider an armature of magnetically permeable material, placed between two pole pieces, as shown in Fig. 8.66. The armature has a coil around it, wound upon a bobbin or encapsulated in plastic, whose magnetic polarity and flux density vary according to the polarity and magnitude of the current which is passed through it. The force attracting the armature towards the north pole is

$$F_N = \frac{K\mu\Phi_1 AH}{d^2}$$

and the force attracting the armature towards the south pole is

$$F_S = \frac{K\mu\Phi_2 AH}{(L-d)^2}$$

The flux between the armature and each pole piece is

$$\Phi_1 = \frac{Ni}{S} \quad \text{and} \quad \Phi_2 = \frac{N(-i)}{S}$$

Combining

$$F_N = \frac{K\mu NiAH}{Sd^2}$$

$$F_S = -\frac{K\mu NiAH}{S(L-d)^2}$$

The force F_T on the armature is the difference between the above forces

$$F_T = F_N - F_S = \frac{K\mu NiAH}{Sd^2} + \frac{K\mu NiAH}{S(L-d)^2} = \frac{K\mu NiAH}{S}\left[\frac{1}{d^2} + \frac{1}{(L-d)^2}\right]$$

When the armature is at the central (null) position, $d = L/2$, so that

$$F_T = \frac{K\mu NiAH}{S}\left[\frac{1}{\frac{L^2}{4}} + \frac{1}{\left(L-\frac{L}{2}\right)^2}\right]$$

$$= \frac{K\mu NiAH}{S}\left[\frac{4}{L^2} + \frac{4}{L^2}\right] = \frac{8K\mu NAH}{SL^2} i$$

where F_T = total force on armature

F_N = force of attraction to north pole

F_S = force of attraction to south pole
Φ_1 = magnetic flux between armature and north pole
Φ_2 = magnetic flux between armature and south pole
A = area of pole pieces
H = field strength of poles
μ = permeability of air
L = distance between pole faces minus thickness of armature
d = distance from north pole face to armature
S = reluctance of iron
N = number of turns in torque-motor coil
i = current through torque-motor coil
K = constant of proportionality

Now, displacement of armature from the null position

$$x = \frac{L}{2} - d \quad \text{so that} \quad d = \frac{L}{2} - x$$

and $\quad L - d = L - \frac{L}{2} + x = \frac{L}{2} + x$

thus $\quad F_T = \frac{K\mu NiAH}{S}\left[\frac{1}{\left(\frac{L}{2}-x\right)^2} + \frac{1}{\left(\frac{L}{2}+x\right)^2}\right]$

$$= \frac{K\mu NiAH}{S\left(\frac{L}{2}-x\right)^2} + \frac{K\mu NiAH}{S\left(\frac{L}{2}+x\right)^2}$$

or $\quad \dfrac{F_T}{i} = \dfrac{F_N}{i} - \dfrac{F_S}{i} = \dfrac{K\mu NAH}{S\left(\frac{L}{2}-x\right)^2} + \dfrac{K\mu NAH}{S\left(\frac{L}{2}+x\right)^2}$

F_N/i and F_S/i are plotted in Fig. 8.67. If F_S/i is subtracted from F_N/i, as shown, it can be seen that F_T/i is almost independent of x, provided that x is small compared with L/2. If, however, the armature is moved too close to either pole piece, this independence is lost.

There are many arrangements that can be used for torque and force motors. The rectilinear solenoid type and the so-called E-I type with the coil wound on the E core are two of many conceivable designs. The arrangement which has gained almost universal use is one in which the magnetic circuit consists of powerful permanent magnets and pole pieces of magnetically permeable material. The armature is mounted on a flexured bearing which has a low spring constant but no friction. The coil is placed around the armature, which may be light in weight.

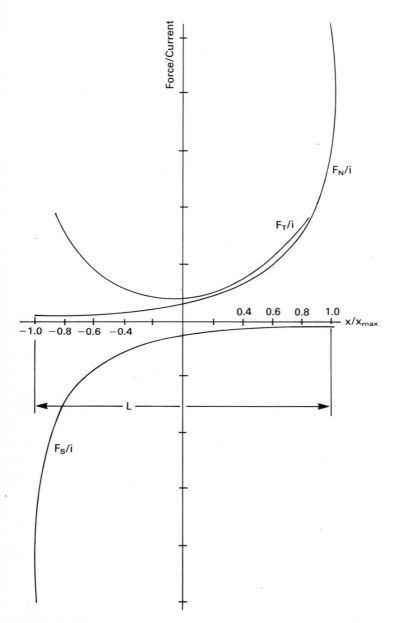

Fig 8.67   Force/current sensitivity vs displacement curves for an electrical torque motor

8.7   Conventional motors

Where an electrical drive having a power rating in excess of (say) 1 kW is required, it will be necessary to use conventional electric motors. It may be necessary to incorporate clutches and/or brakes in order to obtain the correct type of drive: for example, when a drive must develop high holding forces/torques when stationary. Before deciding upon an electric drive, it is advisable to consider the relative merits of alternative hydraulic or pneumatic drives, and a guide to this is presented in Section 9.6.1.

In the case of an AC motor, a conventional polyphase induction motor would normally be used. In order that it can be controlled over a very wide speed range, special control hardware is necessary.

In the past, DC speed control systems have been preferred to AC speed control systems, because of the higher initial costs of the latter (although AC motors are cheaper than the corresponding DC motors). However, the costs of the AC control electronics have dropped to such an extent that both types of system may now be comparable, especially when the improved reliability and maintainability of induction motors, in contrast to commutator-type motors, are taken into consideration.

DC motors can match closely the requirements of the application, because the speed vs torque relationship can be manipulated to almost any useful form, for both motoring and regeneration in either direction of rotation. AC motors stall at torque loads above twice their rating and cannot start loads requiring above about 150% of rated torque; on the other hand, DC motors are often used to deliver momentarily three or more times their rated torque. In emergency situations, DC motors can deliver over five times rated torque without stalling, assuming that the power supply can cope.

Dynamic braking (with which the motor-load energy is fed to a high-dissipation resistor) or regenerative braking (with which the motor-load energy is fed back into the power supply) can be obtained easily with DC motors in applications requiring rapid stopping, thus eliminating the need for, or reducing the size of, a mechanical brake.

DC motor speed can be controlled down to zero, immediately followed by acceleration in the opposite direction, often without switching the power circuit. Because of their high ratio of torque to inertia, DC motors respond relatively quickly to changes in control signal.

DC motors may be classified according to the method of connection of the field windings: separately excited, shunt excited, series excited, and compound excited, the last three types being shown in Fig. 8.68.

Usually, separately excited motors are operated in the same types of configuration as are used with DC servomotors, except that the power rating of the control hardware must be appropriate to the motor size: generally, the field windings will be neither split nor centre-tapped, so that the amplifier power output stage must be chosen accordingly, in the case of field control configurations.

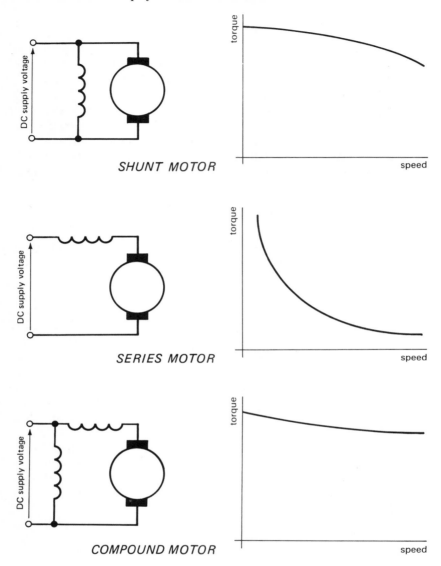

Fig 8.68   Circuits and torque vs speed characteristics for DC motors

8.7.1   DC motor behaviour

Referring to the terms defined by Fig. 8.69, the torque equation for a generalised DC motor can be represented by

$$T = K_T \Phi I_A \, Nm$$

and the EMF equation by

$$E = K_E N\Phi \text{ volts}$$

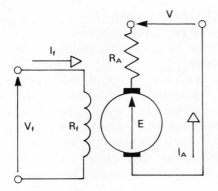

Fig 8.69   Representation of the circuit for a generalised DC machine

In these equations, Φ is nett flux in Wb, T is torque in Nm, N is shaft speed in rad/s; K_T is a torque constant and K_E is an EMF constant. The function of the motor is to convert electrical power into mechanical power, so that this relationship can be presented by the equation

$$TN = EI_A \text{ watts}$$

thus demonstrating that, in the SI system at least, the constants K_T and K_E are numerically equal.

The EMF equation can be rewritten to yield a speed equation

$$N = \frac{E}{K_E \Phi} = \frac{V - I_A R_A}{K_E \Phi} \text{ rad/s}$$

and examination of this equation yields the bases for the alternative techniques available for speed control. Speed can be controlled by manipulating either:

(a)   the armature terminal voltage V, or
(b)   the armature circuit current I_A, or
(c)   the field excitation flux Φ, or
(d)   combinations thereof.

Some of these techniques make use of controlled voltage or current sources. In these cases, the motor would normally be separately excited, and would be operated in one of the following six alternative operating modes:

(a)   manipulated field voltage, constant armature voltage supply
(b)   manipulated field voltage, constant armature current supply
(c)   manipulated armature voltage, constant field excitation
(d)   manipulated armature current, constant field excitation
(e)   manipulated armature voltage, manipulated field excitation
(f)   manipulated armature current, manipulated field excitation.

The principles of operation involved with modes (a) to (d) are precisely the same as those concerned with the control of DC servomotors: these were dealt with in Section 8.6.2. The principles of operation involved with modes (e) and (f) are covered in Section 11.4.1.

All of the techniques alluded to in the previous paragraph incorporate the ability for infinitely variable (un-stepped) speed control of DC motors: operation can migrate across an infinite set of torque vs speed characteristics. A much simpler type of speed control can be effected by operating with a stepped regime, in which only a very limited number of alternative torque vs speed curves is available for selection: this can be established using electromechanical or solid state switches. Resistors can be switched in and out so as to:

(a)   manipulate field current, and therefore excitation flux Φ, in discrete steps, or
(b)   manipulate armature terminal voltage V, in discrete steps, or
(c)   manipulate armature circuit current I_A, in discrete steps, or
(d)   manipulate combinations thereof.

The techniques described above can be adapted to all types of motor excitation: separate, shunt, series and compound. When switching occurs, operation will switch from one torque vs speed curve to another, each of which will be similar in shape to one of those depicted in Fig. 8.68, depending upon the method of excitation employed.

The speed equation shows that speed N can be reversed either by reversing the polarity of the EMF E or by reversing the polarity of the excitation Φ. However, if both E and Φ were to be reversed in sign, no reversal in motor rotation would result: this factor can add to the complexity associated with reversible switched motor drives.

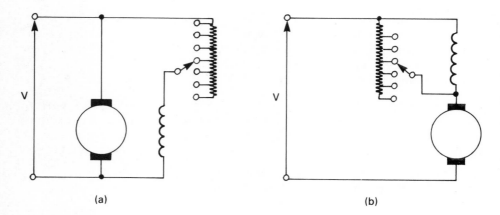

(a)   (b)

Fig 8.70   Use of a switched tapped rheostat to: (a) modify the resistance in series with a shunt field; (b) modify the resistance in parallel with a series field

8.7.2   Stepped field control of DC motors

The motor excitation Φ can be stepped in value by switching values of resistance in and out, either in series with or in parallel with the field circuit. Typical arrangements are shown in Fig. 8.70. The speed achieved with full armature voltage and full excitation is called the 'base speed' of the motor. The switched resistors will always result in a decrease in excitation Φ, so that field control is really useful only for increasing the speed above the base speed value, as is evident from inspection of the speed equation.

Note from the torque equation that, for a given torque T requirement, the current I_A being drawn through the armature must increase as the field is weakened. The resulting potential for increased armature reaction flux concurrently with weakened main flux can create instability, with the result that the motor speed may run away: it follows that this type of speed control must be used with caution.

The potential merits are:

- relative simplicity
- relatively high power efficiency
- relatively low effect upon speed regulation.

The potential disadvantages are:

- speed range is limited to between base speed and (say) 150% of base speed
- potential instability at high speeds, due to armature reaction
- potential commutation problems, due to armature reaction.

8.7.3   Stepped armature resistance control of DC motors

The effective motor armature circuit resistance R_A can be stepped in value by switching values of resistance in and out, in series with the motor armature. A typical arrangement is shown in Fig. 8.71. Reference to the EMF equation shows that increasing the value of R_A above the inherent value for the armature can result only in a decrease in speed below the base value. In this context, the effect is the opposite to that of field control.

Since the main flux Φ is unchanged, there can be no change in armature current I_A if there is no change in torque demand T. However, if the torque demand is increased, there will be a proportional increase in armature current so that, as the EMF equation indicates, the speed will fall.

Therefore, whilst this method of speed control does not suffer from unacceptable armature reaction effects, it can result in poor speed regulation.

Since the series resistor carries the full armature current, it may exhibit high internal power dissipation, so that this method can produce poor power efficiency.

The potential merits are:

- relative simplicity
- the ability to achieve speeds below base speed and, as a consequence
- the ability to combine speed manipulation with motor starting.

The potential disadvantages are:

- high cost of the resistors and switchgear, because of high power and current ratings
- low power efficiency
- poor speed regulation.

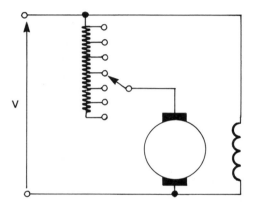

Fig 8.71   Use of a switched tapped rheostat to modify the effective armature circuit resistance R_A

Armature resistance control can be combined with field resistance control, in order to achieve a range of speeds on either side of the base speed. However, this may not be practicable in instances of high motor power ratings where very low speeds are required, because of the very high power dissipation which could result in the armature circuit series resistor. In a case such as this, it may be a practical alternative to effect speed control by connecting resistors in series and in parallel with the motor armature, as shown in Fig. 8.72. Speed is increased by decreasing both R_{se} and R_{sh}. The effect of R_{sh} is to make the speed N less susceptible to changes in load torque T, so that speed regulation is improved, compared with the case where R_{sh} is not present.

The potential merits are:

- improved speed regulation
- the possibility of using R_{sh} for dynamic braking.

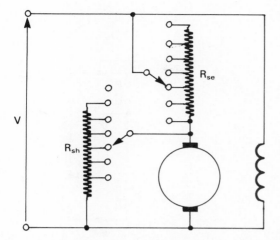

Fig 8.72   Use of a switched and tapped rheostat in series and parallel with the armature

The potential disadvantages are:

- torque reduction resulting from R_{sh} diverting some armature current
- reduced power efficiency, due to extra power dissipation, in R_{sh}.

8.7.4   Reversible switched DC motor drives

The usual method for reversing the direction of rotation of a DC motor is one in which the connections to the armature terminals are reversed, whilst the polarity of the main field is unchanged. In the case of compounded motors, the action of the series field must remain the same relative to that of the shunt field, whenever the armature is switched. Figure 8.73 shows some possible configurations.

8.7.5   Braking of DC motors

A number of alternative switching techniques are available for bringing conventional DC motors rapidly to rest. These may be grouped as follows.

Plugging   The electrical energy to the motor is suddenly reversed, using one of the reversing configurations shown in Fig. 8.73. This causes high deceleration, and a speed-sensing device is used to disconnect the electrical supply when the speed is sensed to have fallen to zero. It may be necessary to switch resistance temporarily into the armature circuit during this phase, in order to limit the armature current to a safe level.

Dynamic braking   The electrical supply to the motor is disconnected, whilst a resistive load is connected across the armature. The mechanical

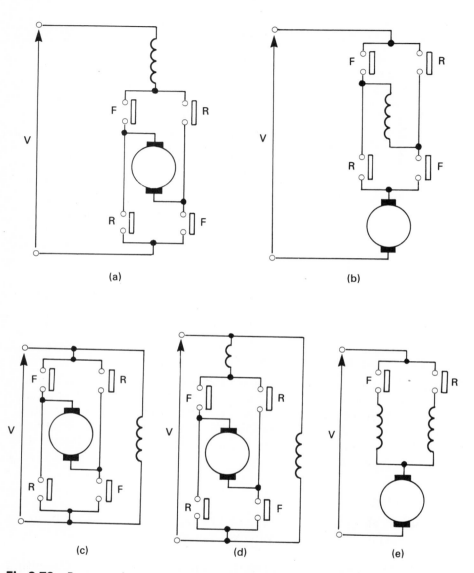

Fig 8.73   Basic configuration for reversible switching of: (a) series motor, armature reversal; (b) series motor, field reversal; (c) shunt motor; (d) compound motor; (e) split series motor. F represents 'forward' and R 'reverse' rotation

energy stored in the mechanical load is converted rapidly, by the motor (which now temporarily behaves as a generator), into electrical energy which is dissipated in the resistive load. The same resistor may be used for the two tasks of starting and braking.

Regenerative braking   This type of braking is used to slow down a motor, but not to bring it to rest. The field excitation is temporarily increased, by using a switch to short out a resistor which is normally present in series with the field. As a result, the EMF E is temporarily greater than the supply voltage V, so that the machine now behaves as a generator: armature current I_A temporarily reverses, causing electrical energy to be transferred to the supply, so that the machine decelerates.

Mechanical braking   See Section 12.6 for information on this topic.

8.7.6 AC drives

AC drives provide accurate speed control, synchronisation of multiple loads, and low maintenance costs. The factors favouring AC drives are:

- rugged, low inertia, reliable motors
- very low maintenance requirements
- insensitivity to severe environments
- safe, convenient, high-speed operation
- 0.5 to 0.25% speed control accuracy with modest cost
- digital speed control accuracy to 0.001%, with increased cost
- ability to phaselock several motors to the same AC supply
- multi motor systems with regeneration are possible.

Most modern variable speed AC motor drive systems employ a variable frequency AC supply, using one of the types of frequency converter described in Section 8.4.5. For motors in the power range from (say) 5 to 150 kW, pulse-width modulated inverters are used: these take DC power, generated from the AC mains by a converter, and invert it to adjustable-frequency AC power for a cage induction motor. Drives may be reversible and may incorporate regenerative braking.

For motors in the power range from (say) 500 kW to 10 MW, cyclo-converters are often used. Drives may be reversible and may incorporate regenerative braking. Typically, a dual cycloconverter is used for each of the three stator phases, with the frequency for continuous operation limited to the range from 0 to 20 Hz.

See Section 11.4.2 for a discussion on general-purpose speed controllers for AC motors.

9

Hydraulic and Pneumatic Amplifiers and Final Control Elements

9.1 Single-stage fluid amplifiers

Fluid pressure or fluid flowrate is the output signal from a fluid amplifier. However, the input signal may be a mechanical displacement, voltage, current or fluid pressure. Fluid amplifiers may be classified as follows.

Liquid amplifiers (usually using hydraulic oil)

- jet pipe
- flapper-nozzle
- spool valve
- Coanda (rarely used).

Gas amplifier (usually pneumatic)

- flapper-nozzle
- Coanda.

9.1.1 Liquid amplifiers

Jet pipe amplifier
   The *jet pipe* amplifier employs a displaceable jet pipe from which high pressure oil is projected at two orifices, to achieve a push-pull type of output. When the pipe is central, the two output pressures will be the same; when the pipe is displaced, one output pressure increases whilst the other diminishes. The magnitudes of the pressures will depend upon the fluid resistance of the load. The internal arrangement of the amplifier is indicated symbolically in Fig. 9.1. The principal application of the jet pipe amplifier is as a preamplifier stage in a two-stage hydraulic servoamplifier.

Flapper-nozzle amplifier
   The internal arrangement of a *flapper-nozzle* amplifier is shown in Fig. 9.2. The nozzle chamber includes a fixed restriction R and a variable

orifice O, the variation being effected by the displacement of the flapper plate by the input signals. When the orifice is fully closed by the flapper, the internal back-pressure (and hence the output pressure p_o) will be equal to the supply pressure p_s; when the orifice is fully open (maximum flapper displacement), the back-pressure will have collapsed to a value approaching sump pressure. The range of flapper displacement, as measured at the nozzle, to achieve the full excursion of output pressure is very small: typically, a displacement of less than 0.01 mm will produce a 5% change in output pressure.

For flapper displacements intermediate between the ends of the range, the device may be visualised as a fluid potential-divider, with an electrical

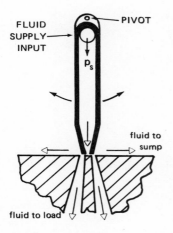

Fig 9.1   Symbolic representation of a jet pipe amplifier

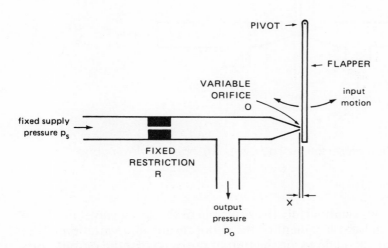

Fig 9.2   Symbolic representation of a flapper-nozzle amplifier

equivalent circuit as shown in Fig. 9.3. In this analog, the fluid resistances are nonlinear, because of the square law relating pressure to flow. A plot of output pressure p_o versus flapper displacement x will be highly non-linear but this characteristic can be linearised by the appropriate application of linear negative feedback. The principal application of the flapper-nozzle amplifier is as a preamplifier stage in a two-stage hydraulic ser-voamplifier.

Where a push-pull type of output is required using flapper-nozzle amplifiers, these may be operated in pairs, arranged as indicated by Fig. 9.4.

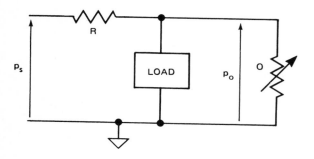

Fig 9.3   Electrical equivalent circuit for a flapper-nozzle amplifier

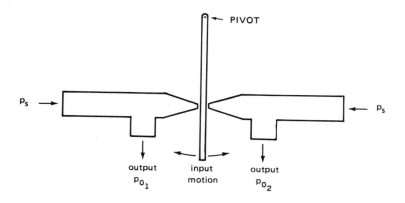

Fig 9.4   Symbolic representation of a push-pull flapper-nozzle amplifier

Spool valve

Figure 9.5 shows symbolically the principle of the spool valve, driving a typical load. Axial displacement of the spool results in a variation in the areas through which oil flows via the output ports, so that the output ports behave as variable orifices. In the example shown, displacement of the

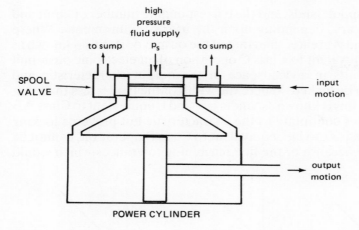

Fig 9.5   A simple spool valve manipulating a double-acting power cylinder

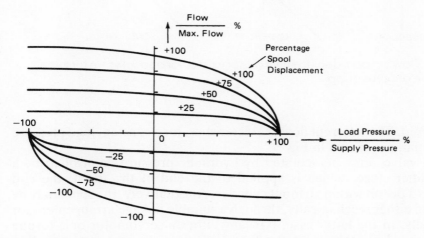

Fig 9.6   Family of flow vs pressure characteristics for a spool valve

spool to the left of its central (null) position will cause supply oil to flow into the left-hand side of the power cylinder: this will drive the power piston to the right and oil from the right-hand side of the power cylinder will be exhausted to the sump via the right-hand output port of the spool valve. Displacement of the spool to the right will similarly cause the piston to be driven to the left.

   Typically, the excursion range of the spool will be very small, in the order of 1 to 2 mm overall. Because of the square law relationship for orifices, the load flow versus load pressure characteristic for the spool valve is highly nonlinear, as exemplified by Fig. 9.6.

The number of spool 'lands' and the corresponding number of input and output ports will vary, depending upon the user's requirements. Where the width of the land is greater than that of the output port, this is known as 'overlap' and is equivalent to Class C operation of an electronic push-pull amplifier: this results in a deadspace in the output characteristic and enables the load to be locked. Where the width of the land is less than that of the output port, this is known as 'underlap' and is equivalent to Class AB operation: this gives continuity to the characteristic but prevents locking from being achieved. Zero lap, equivalent to Class B operation, cannot be achieved precisely because of the fine machining tolerances which would

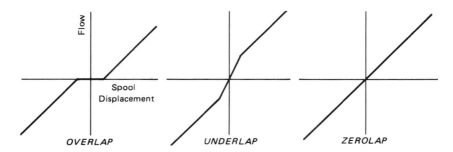

Fig 9.7   Flow vs spool displacement characteristics for a spool valve for three alternative degrees of lap

be required. The different flow characteristics, for a constant load pressure differential, are shown in Fig. 9.7.

Spool valves may be actuated by hydraulic pressure, controlled by a preamplifier stage, which is applied to the end of the spool: the end becomes a piston within an input cylinder. Alternatively, the spool may be displaced either mechanically, through a suitable linkage arrangement, or electrically: in the latter case, typically either a servomotor or a torque motor would be suitably geared to the spool, so as to have its rotary motion converted to rectilinear motion. Another electrical alternative is to use a force motor, directly connected mechanically. Occasionally, the spool is 'loaded' with compression springs, in order to provide it with the required displacement versus force characteristic.

Negative feedback is often applied around a spool valve, with pilot amplifier stages and/or the load sometimes being included within the loop. The feedback will linearise the characteristic of the valve and, if appropriately configured, may be used to convert the valve from a flow source into a pressure source. The feedback signal may be applied directly to the spool or sometimes to a ported sleeve which is interposed between the spool and the valve cylinder. The feedback signal may originate either as a mechanical displacement, transmitted through linkages, a hydraulic pressure, or as an electrical signal from a suitable transducer.

Coanda amplifier

A jet of fluid emerging from a nozzle tends to deflect towards an adjacent surface and, under certain conditions, will attach to it. This phenomenon is known as the 'Coanda' or 'wall attachment' effect. The presence of an adjacent surface creates an area of turbulence and low pressure on one side of the jet. The imbalance of pressure forces across the jet causes it to bend away from its free-flowing direction.

Although the Coanda effect is well known in pneumatic devices, it is rarely used when the working substance is a liquid. However, Fig. 9.8 shows a novel use of the Coanda effect for liquid level control. Liquid entering the valve is formed into a jet by a nozzle at the valve inlet. This jet passes by the control port and through a steering chamber, at the end of which are two outlets. The pressure developed in the control port acts on one side of the jet. When the level falls below the end of the dip tube, the valve diverts liquid to the tank. When the level reaches the dip tube, liquid is diverted to a tank by-pass line. The entire flow can be diverted to one outlet, although the outlet may be at a higher pressure than that at the other.

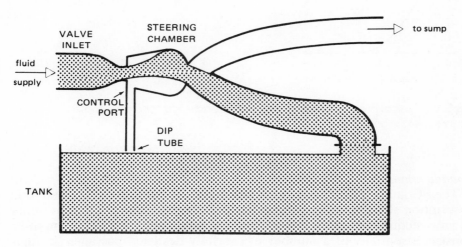

Fig 9.8 A Coanda amplifier used to control liquid level in a vessel

9.1.2 Gas amplifiers

Flapper-nozzle amplifier

Most pneumatic amplifiers employ the flapper-nozzle principle. This is because compressed air is not lubricating (unless oil mist is injected), it does not have the same heat-removing capability as oil, and it tends to include small particles of foreign matter, all of which render spool valves unsuitable for pneumatic service: the flapper-nozzle has a high tolerance to these disadvantages.

Figure 9.1 and the description in Section 9.1.1 also applies when compressed air is the operating medium, except that the air now exhausts to atmosphere (creating a characteristic hissing noise) and that the flapper excursion is now typically in the order of less than 0.1 mm. Figure 9.9 shows a typical pressure vs displacement characteristic for a pneumatic flapper-nozzle amplifier.

Because of its high sensitivity, this type of amplifier may be regarded as the pneumatic equivalent of the electronic operational amplifier and, as a result, it is used in a number of analog computing types of application.

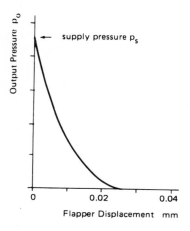

Fig 9.9 Typical pressure vs displacement characteristic for a pneumatic flapper-nozzle amplifier

Coanda amplifier

The 'Coanda effect' has been described in Section 9.1.1, and this description also applies in pneumatic applications. Figure 9.10 represents a cross-section through a typical fluidic Coanda amplifier. The amplifier consists essentially of a number of carefully designed passages through which fluid may flow. If the velocity of the fluid emerging from the nozzle is such that a critical value of Reynolds number is exceeded, the stream will attach to one wall, due to the Coanda effect, and exit via one output port. If a jet of fluid is now injected into the appropriate control port, the pressure in the separation bubble will rise and the main jet will flip over and attach to the other wall. It is found that the energy required to be injected in the control port is considerably less than the main jet energy, and hence amplification has been achieved.

Air relay

The *air relay* is the pneumatic equivalent to the electronic unity follower power amplifier. Normally, it has a pressure gain of approximately unity

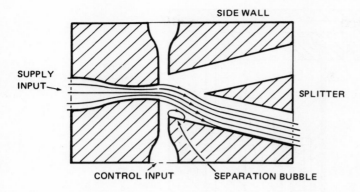

Fig 9.10   Cross-section through a typical pneumatic Coanda amplifier

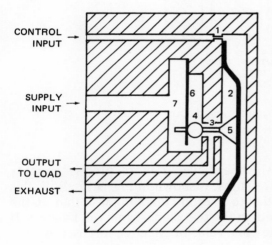

Fig 9.11   Cross-section through a typical open-loop air relay

but the output flowrate may be many times the input flowrate which the amplifier draws. The relay is particularly suitable as the output stage of a multi-stage pneumatic amplifier. Figure 9.11 shows a cross-section through one type of air relay. The input pressure enters through port (1) and is applied to the right-hand face of the diaphragm (2). The motion of the diaphragm to the left, against the restraining force of the spring (6), will move the valve stem (3) to the left. As a result, the ball valve (4) will open further, whilst the opening of the exhaust valve (5) will be reduced. The supply air in compartment (7) will bleed through the two valves and exit through the exhaust port. The pressure at the output port will assume an intermediate value which will increase towards the supply pressure as the valve stem moves to the left: the two valves behave as a pneumatic (nonlinear) potential divider.

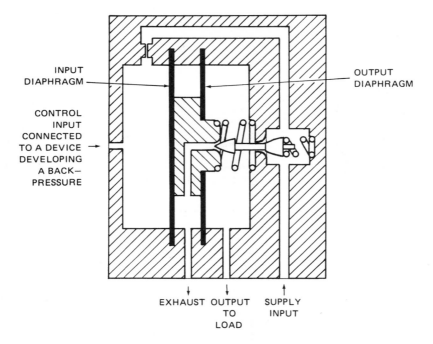

INPUT DIAPHRAGM

OUTPUT DIAPHRAGM

CONTROL INPUT CONNECTED TO A DEVICE DEVELOPING A BACK-PRESSURE →

EXHAUST OUTPUT TO LOAD SUPPLY INPUT

Fig 9.12 Cross-section through a typical closed-loop air relay

A more complex version, which achieves a more precise equality between the input and output pressures, is shown in Fig. 9.12. With this type, the output pressure is applied to a second diaphragm so that, in the steady state, the forces developed on the two diaphragms will balance. If these two diaphragms present the same value of effective surface area, the output pressure will match accurately the input pressure.

9.2 Multistage fluid amplifiers with feedback

In general, the disadvantages with single-stage fluid amplifiers will be as follows:

- the characteristics are highly nonlinear
- although the output pressure range may be adequate, the flowrate limit may be far too small for the load
- the dynamic behaviour may be too slow
- the accuracy may be inadequate.

A considerable improvement can be effected using multistage amplifiers with linear negative feedback. Some techniques were discussed in Section 9.1.1. By multistaging hydraulic valves, a power gain in excess of 10^5 can

typically be achieved, with excellent steady state and dynamic performance.

9.2.1 Servovalves

Figure 9.13 is an example of an electrical force motor actuating the flapper
of a push-pull pair of flapper-nozzle amplifiers which in turn drive, by
means of their output pressures, the spool of a spool valve. Feedback from
the spool displacement to the flapper plate is provided by the feedback leaf
spring and hydraulic pressure feedback from the spool valve output is
applied to the spool, so that this three-stage amplifier has two feedback
paths.

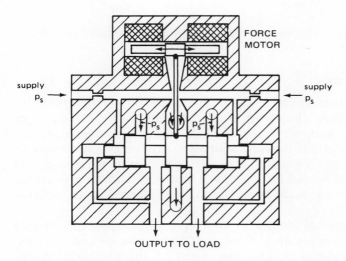

Fig 9.13 Cross-section through a servovalve incorporating three stages of
amplification and two feedback paths

9.2.2 Pneumatic amplifiers

Figure 9.14 is an example of a two-stage pneumatic amplifier. The flapper
plate will assume a quiescent position such that the input force is balanced
by the feedback force arising from the output pressure within the feedback
bellows. Thus, the output pressure will be proportional to the input force,
irrespective of nonlinearities within the amplifier stages. The outflow
which can be delivered to the load is much greater than that which the
flapper-nozzle amplifier could supply.

The system of Fig. 9.14 may also be used as a pneumatic displacement
transducer since, taking into account the spring rate of the feedback
bellows, a displacement balance will also exist at the flapper plate, in the
steady state.

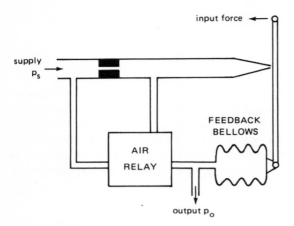

Fig 9.14   Symbolic representation of a typical two-stage pneumatic amplifier

9.3   Hydraulic pumps

Hydraulic pumps are usually employed to provide the flow in high-pressure hydraulic power supplies, when normally they would be driven at constant speed, typically by an induction motor. However, gear and vane pumps may also be used as variable flowrate sources in process pipelines and, in this type of application, they would typically be driven by variable-speed electric motors: there would be an almost linear relationship between flowrate and shaft speed. Such pumps are often referred to as 'positive displacement pumps'.

There also is a range of axial and rotary piston pumps which would typically be used to drive piston motors: such a pump-motor combination can be regarded as the hydraulic equivalent to the electrical Ward-Leonard system of Section 8.4.1. With these pumps, there is an almost linear relationship between outflow rate and the input signal, with the pump shaft being driven at constant speed typically by an induction motor. The input signal is in the form of a displacement, and this may be derived from a mechanical linkage, an electrical torque motor, or a pneumatic or hydraulic actuator.

9.3.1   Gear pumps

Figure 9.15 shows a cross-section through a simple two-pinion gear pump. One gear is driven by the prime mover and, in turn, drives the other gear. The two gears draw oil in through the inlet port, propel it around the walls of the case, and expel it through the outlet port. The volumetric flow rate will be proportional to the gear shaft speed.

These pumps are used for pressures up to about 2000 psi (13.8 MPa). They are low-cost items, mechanically simple, and reliable but they exhibit relatively low power efficiency.

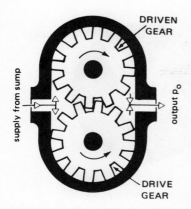

Fig 9.15 Cross-section through a typical two-pinion gear pump

9.3.2 Vane pumps

Figure 9.16 shows a cross-section through a simple unbalanced sliding vane pump. The vanes are spring loaded and are free to slide radially in and out of slots in the eccentrically mounted rotor. In the drawing shown, cavities passing the inlet port are expanding in volume, until they pass the uppermost point on the case, after which they progressively contract in volume as the outlet port is approached and passed. Thus, a suction action will be created on the inlet side, drawing oil in, and a compression action will be created on the outlet side, propelling oil out. The volumetric flow rate will be proportional to the rotor shaft speed.

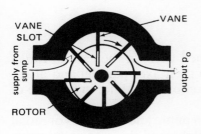

Fig 9.16 Cross-section through a typical unbalanced vane pump

These pumps are used for pressures up to 1500 psi (10.3 MPa). They are more costly, mechanically less simple, and less reliable than gear pumps but they exhibit medium levels of power efficiency.

9.3.3 Piston pumps

Piston pumps consist of a symmetrical set of cylinders with their pistons interconnected, together with an appropriate rotating set of valve gear.

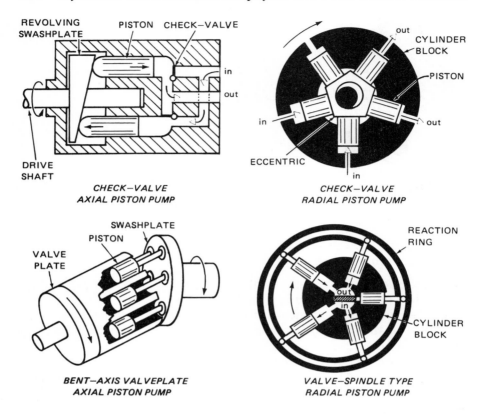

Fig 9.17 Two types of axial piston pump and two types of radial piston pump

The pistons may radiate with radially aligned axes, in the case of *radial piston pumps*, or they may be aligned with their axes parallel and distributed around a circle concentric with the pump centre line, in the case of *axial piston pumps*.

Figure 9.17 shows two types of axial piston pump and two types of radial piston pump. In most cases, the cylinder block and the associated valve gear are rotating, driven through the input drive shaft by a prime mover. At all instants in time, half the pistons are being withdrawn from their cylinders, whilst the other half of the pistons are being driven into their cylinders. The valve gear is organised so that those cylinders in which the enclosed volume is expanding are connected to the inlet port, whilst those cylinders in which the enclosed volume is contracting are connected to the outlet port. Thus, suction will be established at the inlet and compression will be established at the outlet.

In many of these pumps, the length of the piston stroke may be varied. In the case of the radial pumps, this is effected by varying the eccentricity of the rotor axis relative to the axis of the case. In the case of the axial pumps, the angle of inclination of the swashplate face, relative to the axis of the cylinder block, can be varied appropriately. The direction of inflow and outflow can be reversed, if the eccentricity or tilt-angle can be reversed, relative to the null position (at which the stroke is zero).

The volumetric flow rate will be proportional to the product of the drive shaft speed (usually constant) and the piston stroke (often variable). These pumps are used for pressures in the approximate range of 1500 to 5000 psi (10.3 to 34.5 MPa). They are expensive, mechanically complex, and have lower reliability than other pumps but they exhibit high levels of power efficiency.

9.4   Final control elements

Final control elements for fluid amplifiers are mainly linear actuators (cylinders, rams, jacks), for rectilinear motion, and rotary actuators and motors for rotary movement.

9.4.1   Linear actuators

The construction of hydraulic cylinders differs from that for pneumatic cylinders, but the principles of operation are the same. Hydraulic cylinders are much more bulky, because of the much higher pressures involved. Hydraulic systems may operate up to 5000 psi (34.5 MPa) whereas pneumatic systems rarely operate much above 100 psi (690 kPa), at the final control element. Hydraulic cylinders often include 'cushioning', to minimise the impact of the piston on the end of the cylinder: this requires a small chamber, at the end of the cylinder, in which oil becomes trapped by the end of the piston and compressed as the piston approaches the limit of travel. Table 9.1 lists some alternative styles of cylinder construction.

A 'jack' is a self-contained unit, consisting of a cylinder, control valve, and piping.

Pressure applied to a piston produces a piston velocity proportional to the volumetric inflow rate, if compressibility and leakage effects are ignored. The force required will depend upon the characteristics of the mechanical load, and the internal pressure developed will be equal to this force divided by the effective area of the piston face. In the case of pneumatic and high-performance hydraulic systems, compressibility effects must be taken into account.

9.4.2   Rotary actuators

Where rotary motion with limited angular travel is required, a rotary actuator may be used. Alternatively, it would be possible to use a linear actuator in combination with a rack and pinion. Figure 9.18 illustrates three alternative styles of rotary actuator.

Table 9.1   Alternative styles of cylinder construction

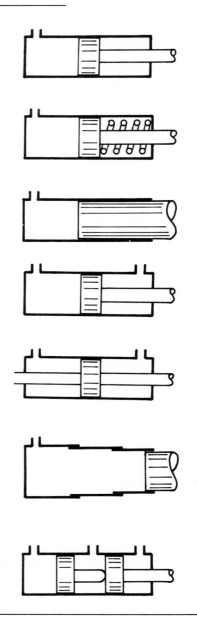

Single-Acting
Power stroke in one direction only, return being effected by some external means.

Spring Return
Single-acting cylinder with return stroke effected by a spring. Length of cylinder is at least twice actual stroke length, because of spring length.

Ram
Single-acting cylinder with rod diameter equal to the piston diameter and to the bore of the cylinder.

Double-Acting
Cylinder with power stroke in both directions. Actuating fluid lines are connected to both heads of the cylinder, usually through a spool valve.

Double-End Rod
Cylinder with rod extending from both sides of piston. Usually double-acting, such a cylinder requires a rod bearing and packing in each cylinder head.

Telescopic
Collapsed length is less than one half of extended length. Cylinder is made of two or more hollow sections which telescope together in the retracted position. Telescopic cylinders usually are single-acting.

Positional
Stroke is split up into two or more segments. Piston rod can be actuated to any of a set of alternative positions.

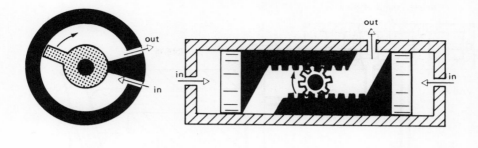

VANE TYPE ACTUATOR   PISTON–AND–RACK ACTUATOR

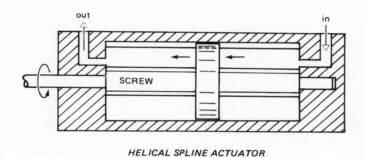

HELICAL SPLINE ACTUATOR

Fig 9.18 Three alternative types of rotary actuator

9.4.3 Hydraulic motors

The gear pumps of Section 9.3.1 and the vane pumps of Section 9.3.2 may be operated in reverse, with hydraulic fluid now the input and shaft rotation as the output. These devices now become *gear motors* and *vane motors*. However, because they possess relatively low efficiencies and poor low-speed characteristics, it is much more common to use piston types of hydraulic motor, where continuous rotary motion is required.

The pumps illustrated in Fig. 9.17 may be operated in reverse, thus becoming *radial piston motors* and *axial piston motors*. The eccentricity or the tilt angle between the face of the swashplate and the axis of the cylinder block (as the case may be) is fixed permanently at one value. At any one time, half the pistons are being forced outwards by pressurised oil flowing into the motor, whilst the other half of the pistons are being drawn inwards due to the suction action of oil flowing out of the motor. The combination of the translational forces in the set of piston rods will react to produce angular motion of the cylinder block. The direction of angular motion will reverse when the direction of oil flow is reversed, and the speed of rotation

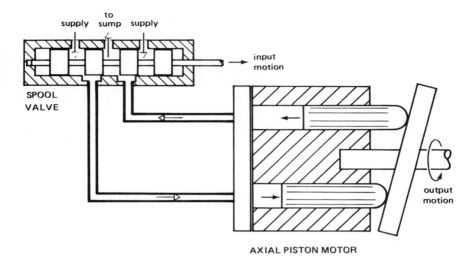

Fig 9.19   Typical hydraulic drive using a spool-type servovalve manipulating a swashplate motor

will be proportional to the volumetric flowrate, if compressibility and leakage effects are neglected.

Hydraulic motors may be controlled either by servovalves or by hydraulic pumps arranged to deliver variable flowrate. Figure 9.19 shows a typical installation of a spool-type servovalve being used to drive a swashplate axial piston motor.

9.5   Block diagrams for hydraulic drives

Figure 9.20 shows a small-signal block diagram of the drive represented by Fig. 9.19. In this:

V = voltage applied to the force motor (V)
I = current in force motor coil (A)
Q = oil flowrate between valve and motor (l/s)
Q_l = leakage flowrate, past pistons (l/s)
Q_c = equivalent compressibility flowrate (l/s)
Q_m = useful flowrate to motor (l/s)
p = differential pressure across the motor (Pa)
T = nett torque developed by the motor (Nm)
N = angular velocity of the motor (rad/s)
R_m = resistance of force motor coil (Ω)
T_m = electrical time constant of force motor coil (s)
T_v = mechanical time constant of valve (s)
$\partial Q/\partial I$ = sensitivity of flowrate to changes in force motor current (l/s per A)

$\partial Q/\partial p$ = sensitivity of flowrate to changes in differential pressure across the motor (l/s per Pa)

K_t = motor torque sensitivity (Nm/Pa)

K_q = motor flow sensitivity (l/rad)

V   = volume of oil in motor and pipes (l)

B   = bulk modulus of oil (Pa)

L   = leakage coefficient (l/s per Pa)

The bulk modulus is defined in order to accommodate any compression and expansion of the fluid, entrained vapour or air, and the pipe walls which may occur under pressure. The diagram does not include any minor negative feedback paths, which may be incorporated for linearisation and other purposes.

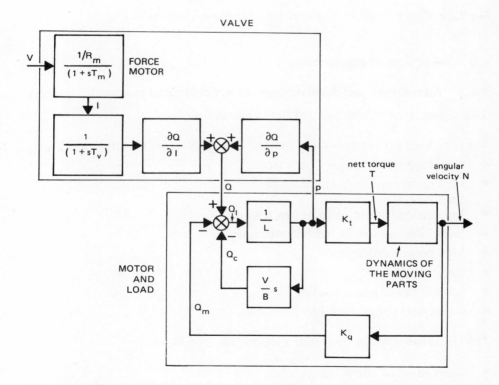

Fig 9.20   Block diagram representation of the hydraulic drive shown in Fig 9.19

Figure 9.21 shows a block diagram for a power cylinder and load: typically, the cylinder would be controlled by a servovalve. This diagram is similar to the representation, in Fig. 9.20, of the motor and load, except that the constant A has been introduced to represent the effective surface area of the piston face (in m^2).

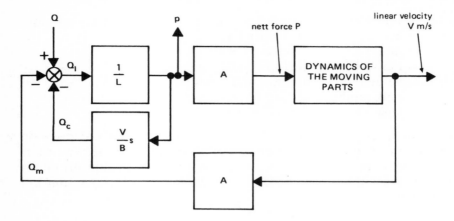

Fig 9.21   Block diagram representation of a power cylinder and load

9.6   Selection of equipment

9.6.1   Advantages and disadvantages of hydraulic and pneumatic systems

The advantages of hydraulic systems are as follows:

- very high power-to-weight ratio
- very high peak torques (or forces)
- very short time lags enabling fast response

} { A direct result of operating at high pressure (up to 5000 psi, 34.5 MPa).

- smooth operation even at low speeds
- large holding forces, when stationary
- high power efficiency
- rugged
- self lubricating and cooling
- linear and rotary motions available.

The disadvantages of hydraulic systems are as follows:

- clumsiness of connections
- mess from oil leaks
- danger from large oil leaks (high pressure, toxicity, and flammability)
- oil properties change with temperature
- inertia of oil can degrade system performance
- expensive when power supply is included in cost
- fluid must be kept clean and gas free
- severe nonlinearities may necessitate special treatment.

The advantages of pneumatic systems are as follows:

- relatively cheap
- pressurised air is often available, so that no special supply is then required
- safe in potentially explosive atmospheres
- no mess or danger from leakage
- can exhaust to atmosphere.

The disadvantages of pneumatic systems are as follows:

- power-to-weight ratio is lower than with hydraulics (due to the lower operating pressures)
- relatively inefficient
- undesirable behaviour due to compressibility of air — for this reason, pneumatic drives tend to be used mainly in on-off (bang-bang) applications.

9.6.2 Selection of hydraulic servovalves and final control elements

In selecting a servovalve, several questions must be considered.

- Is it flow or pressure that is to be controlled?
- Is the intention to control the velocity, position, or force developed by the final control element?
- Is gain compensation required?
- What frequency response is needed?
- How large must the valve be?
- Is three-way or four-way porting needed?

In addition to this, the size of the final control element must be determined. The usual procedure is to plot a pressure versus flow curve for a given final control element and load. Over this plot is superimposed the servovalve performance capability, which must be slightly outside the load curve for satisfactory overall performance.

9.7 Power supplies

When hydraulic or pneumatic components are used, it will be necessary to provide a source of pressurised oil or air for them. Many plants have high volume compressed air mains, typically at a pressure in the region of 100 psi (690 kPa): this may be suitable for the final control elements but most pneumatic instruments and amplifiers require air supplies in the 20 to 25 psi (138 to 172 kPa) pressure range. The lower pressure supply is easily derived from the higher pressure supply, using a pressure-reducing valve having an adequate flow capacity. In the case of hydraulic systems, it is necessary usually to construct a hydraulic power supply especially for these systems.

9.7.1   Hydraulic power supplies

Figure 9.2.2 shows a hydraulic circuit diagram for a typical hydraulic power supply, together with an electrical equivalent circuit. The positive displacement pump is driven at constant speed, by the induction motor, and behaves as a constant flowrate source. The pressure relief valve returns oil to the sump whenever the developed pressure attempts to

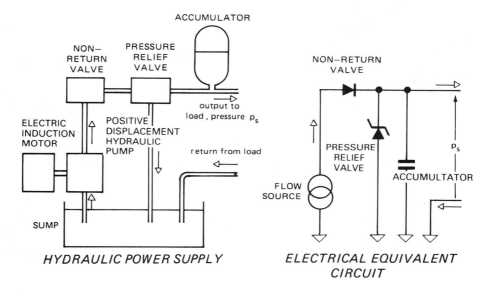

Fig 9.22   Hydraulic circuit diagram and electrical equivalent circuit for a hydraulic power supply

exceed the value manually set into the valve. The non-return valve ensures that the load cannot back-drive fluid to the pump during transient operation.

The accumulator behaves as a hydraulic capacitor and smooths out any short-term fluctuations in output pressure. It can take a number of different forms, some of which are shown in Fig. 9.23. All operate on the principle that the accumulator stores the potential energy of the oil, which is being held under pressure by an external force, acting against the forces arising from the dynamic behaviour of the hydraulic system.

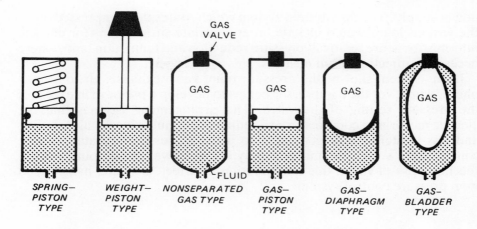

Fig 9.23   Cross-sections through six alternative types of hydraulic accumulator

9.7.2   Pneumatic power supplies

Figure 9.24 shows a typical installation for a pneumatic power supply, where this has to be provided.

The necessary components, taken in sequence, typically would be:

1 air intake with particle filter
2 air silencer, to minimise acoustic noise
3 air compressor (a pneumatic piston pump) driven at constant speed by an induction motor
4 pressure relief valve, to determine output pressure
5 cooling heat-exchangers
6 moisture separator, to remove water which has condensed during the compression process
7 air receiver, which is a storage cylinder and is equivalent to the accumulator in the hydraulic system.

The electrical equivalent circuit would be similar to that for the hydraulic

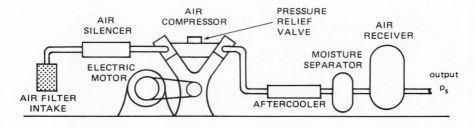

Fig 9.24   Typical installation for a pneumatic power supply

power supply. The distribution system which routes the compressed air to the various loads would include, at appropriate sites, means for particle filtering, moisture removal, pressure reduction and regulation, and, where necessary, injection of oil mist.

In the system shown, the pressure relief valve exhausts air to atmosphere, whenever the compressor begins to develop a pressure in excess of the pressure setting of the valve. This arrangement assumes that the electric motor and compressor are running continuously. An alternative, and less wasteful, arrangement is to instal a pressure-actuated switch, connected to switch the motor supply off whenever the output pressure reaches a preset value: this configuration, in effect, is a bang-bang closed loop pressure control system.

10

Flow Control Valves, Actuators and Positioners; Pneumatic Process Controllers

10.1   Flow control valves

In many process control loops, control of the system controlled variable is effected by manipulating the flowrate of a fluid in a pipeline. Where sufficient pressure head exists at the source of the flow, the rate of flow may be manipulated by inserting a variable, controllable restriction: that is, a fluid resistance. A flow control valve performs this task and is, in effect, an adjustable orifice.

There are many different styles of valve made for throttling the flow of liquids, slurries, gases and vapours, and they may be categorised as follows.

Valves having a rectilinear motion

- single and double seat globe
- split body globe
- three-way globe
- angle
- diaphragm
- needle
- pinch
- gate
- wedge
- slide

Valves having a (limited) rotary motion

- butterfly
- ball
- plug

Figure 10.1 shows the principles of operation of a number of these valves. Many valves installed in process plants will be operated manually, but in this volume we are concerned with those versions which can be actuated by pneumatic, electrical or hydraulic means. All of the valve types listed above can be configured to be signal actuated, either in a fully activated/ inactivated mode or in an infinitely variable mode. In practice, however, the control valves most commonly installed for automatic control use are the single and double seat globe types: further discussion will be restricted almost exclusively to these.

The construction of the globe valve may be separated into three principal parts.

Actuator   This is the source of motive power to position the valve plug against the reaction of fluid forces on the plug, weight of the moving parts, and any spring force which may be introduced.

Body   This is the case for the variable orifice, is usually a die-casting, and contains the valve trim.

Trim   This is the combination of the fixed valve seat and the moving valve plug, the relative position of which determines the area of the adjustable orifice.

Figure 10.2 shows a cross-section through a typical double-seated control valve. The advantage with the double seat is that the reaction forces resulting from the pressure and flow of the fluid tend to cancel, so that the force requirements for the actuator are low; however, if a tight closure is desired, this is achieved more readily with a single-seated valve. Figure 10.3 shows alternative seat arrangements.

The style, size, and material of the body and trim have to take into account:

- the maximum operating pressure in the fluid
- the nature of the fluid, in terms of corrosive, toxic, and flammable properties
- the possibility of entrained solids being present in the fluid and of the building up of solid residues
- the possibility of the occurrence of 'cavitation', which arises from pressure recovery causing bubbles of vapour or gas to be reabsorbed spontaneously, in a liquid flow medium, by a process of implosion
- the possibility of the occurrence of 'flashing', which arises from pressure reduction causing the generation of bubbles of vapour or gas previously dissolved in a liquid flow medium
- the type of installation
- the properties of the fluid, such as the fluid phase, density, viscosity and vapour pressure

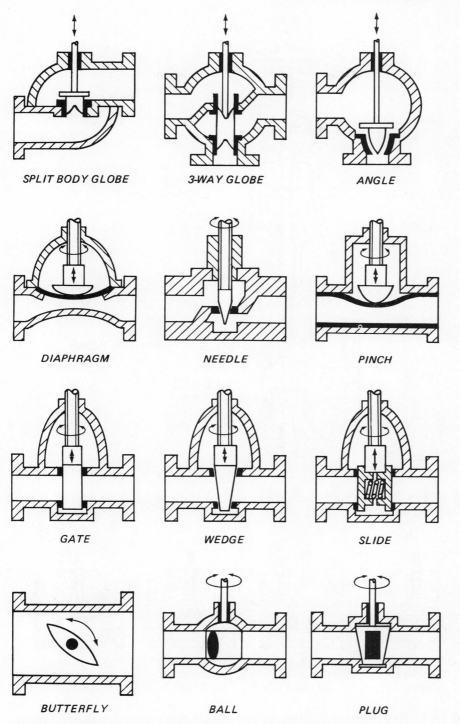

Fig 10.1   Cross-sections through twelve different types of control valve, showing the principles of operation

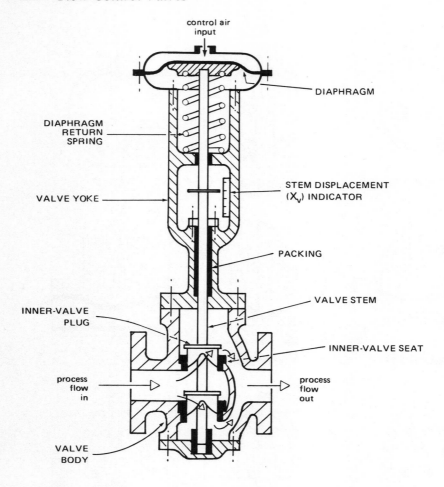

Fig 10.2   Cross-sections through a typical double-seated globe type of control valve

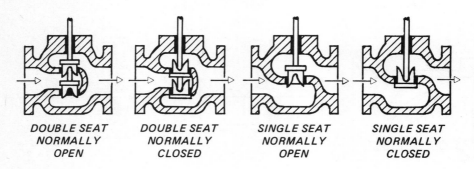

Fig 10.3   Four alternative plug and seat arrangements for a globe type of inner valve

- the range of flowrates required
- dynamic performance requirements.

The engineering procedure for determining the correct size of a valve body is referred to as 'control valve sizing', and this has been rationalised by use of a 'valve flow coefficient', denoted by C_v. The definition of C_v may vary between manufacturers, but a common one is the number of US gallons of water which will flow through the wide-open valve in one minute, when the pressure drop across the valve is 1 psi. The corresponding flow formula for a liquid flow medium would be

$$Q = C_v \sqrt{\frac{\Delta p}{G}}$$

where

Q = volumetric flowrate through the fully open valve, in US gal/min
Δp = pressure drop across the valve, in psi
G = specific gravity of the liquid

In the SI system of units, the equivalent flow formula takes the form

$$Q = A_v \sqrt{\Delta p/\rho}$$

where

A_v = valve flow coefficient
Q = volumetric flowrate through the fully open valve, in m^3/s
Δp = pressure drop across the valve, in Pa
ρ = density of the liquid, in kg/m^3

10.1.1 Valve inherent characteristic

The *inherent characteristic* of a control valve is the relationship between the flowrate Q through the valve and the valve position X_v (the displacement of the valve plug), with a constant pressure drop across the valve. Since the pressure drop across the valve will not be constant when the valve is in service, the inherent characteristic must be measured under specific test conditions. The shape of the characteristic is dependent upon the contouring of the valve plug, and examples of different contouring are shown in Fig. 10.4.

Figure 10.5 shows the forms of inherent characteristic which can be obtained for representative plug types. The *quick opening* characteristic provides a large change in flowrate for a small change in valve position, and is used in on-off applications requiring rapid opening or closure. The *linear* characteristic is normally used in systems in which, when the valve is installed, most of the system pressure is dropped across the valve. The *equal percentage* characteristic is one having a gradient proportional to the flowrate. It is used typically in applications where only a small proportion

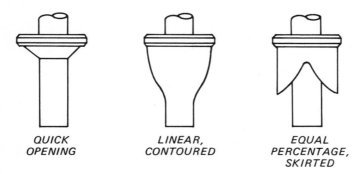

QUICK   LINEAR,   EQUAL
OPENING   CONTOURED   PERCENTAGE,
SKIRTED

Fig 10.4   Three alternative contours for a valve plug

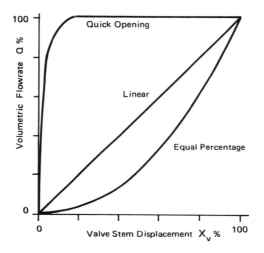

Fig 10.5   Three alternative valve inherent characteristics

of the system pressure is dropped across the valve. The inherent characteristic, with a given plug profile, may be modified by using a valve positioner (see Section 10.1.4) having a nonlinear feedback, by virtue of the use of a suitably profiled cam in the positioner feedback path. Alternatively, the static characteristic of the controller driving the valve actuator may be modified appropriately.

10.1.2   Valve installed characteristic

The electrical analog of the control valve is a variable nonlinear resistance in the steady state. When installed in a flow line, the complete line would resemble a set of series-connected nonlinear resistances, as shown in Fig. 10.6, each one representing a pipeline component presenting a resistance to flow. (The nonlinearity occurs because pressure drop is proportional to the square of the flowrate). The *installed characteristic* will be the relationship between the installed flowrate Q and the valve position X_v, when the supply pressure is constant: this will obviously be dependent upon the combination of the inherent characteristic and the properties of the other components in the flow line.

Except in on-off applications, the most desirable installed characteristic is usually a proportional (linear) relationship. Note, however, that the process engineer will need to supply, to the valve manufacturer, his inherent characteristic requirements and obviously these will be affected by the nature of the installation.

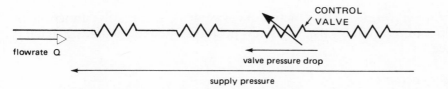

Fig 10.6 Steady state electrical equivalent circuit for a pipeline containing a control valve

10.1.3 Control valve actuators

The requirements for a valve actuator depend upon the type of valve being controlled. For example, a globe type valve will require rectilinear motion, whereas a butterfly valve will require rotary motion. Each type of actuator possesses certain merits, in relation to the others.

Factors involved in actuator selection

- The actuator must be compatible with the controller or valve positioner, depending upon which is the signal source.
- The actuator must generate sufficient force to overcome the reaction forces imposed by the valve.
- The actuator must be appropriate to the operating environment: for example, pneumatic actuators would be preferred in potentially explosive atmospheres or in environments requiring high levels of hygiene.
- The actuator stroke must match the requirements of the valve.
- The actuator must meet the speed-of-response requirements.
- The actuator must leave the valve in the most desirable state, should the power supply fail.

Sources of reaction force
   The causes of components of reaction force, to be overcome by the actuator, are as follows.

Weight and inertia   These are the unbalanced weight and associated inertia of all the moving parts of the valve and its actuator. Except for very large valves, the inertia forces are not significantly large.

Friction   This is due to the action of the sealing gland on the valve stem and also sometimes to the actuator itself. This is a combination of stiction and coulomb friction, and contributes to mechanical hysteresis and hence positioning accuracy of the valve.

Fluid pressure   This is the pressure of the fluid on the valve plug and the associated force can be diminished considerably by the use of double-seated arrangements.

Diaphragm spring   Many actuators include a spring, the function of which is to return the valve to its new position when the actuating signal is reduced.

Types of actuator

Actuators may be categorised as follows.

- pneumatically operated diaphragm actuators
- piston actuators, pneumatically or hydraulically operated
- electrohydraulic actuators
- high-performance servo actuators
- electromechanical actuators
- manually operated handwheel actuators.

By far the most common variety of actuator is the pneumatically operated diaphragm type. This consists of a synthetic rubber diaphragm sandwiched between the flanges of two circular pressed-steel dishes, as shown in Fig. 10.2, in cross-section. The 3 to 15 psi (20 to 100 kPa) control air signal may be applied either to the top or the bottom face of the diaphragm, depending upon requirements; a restoring force will be provided by a calibrated spring placed in the other chamber (that is, the one not supplied with control air). Various combinations of actuating action and valve action (which depends upon the orientation of the valve trim) are illustrated in Fig. 10.7. A cast yoke attaches the actuator to the valve body, as shown in Fig. 10.2.

The cylinder of a piston actuator is made from a casting or a pressing and can withstand much higher control pressures than can the diaphragm type. For this reason, piston actuators are particularly useful in applications where the pipeline pressure is high. The cylinder is mounted, with the piston axis aligned with the valve stem, vertically above the valve yoke. Pneumatic piston actuators are normally used in combination with valve positioners, which are described in Section 10.1.4.

With a hydraulic piston actuator, the piston may be controlled by an

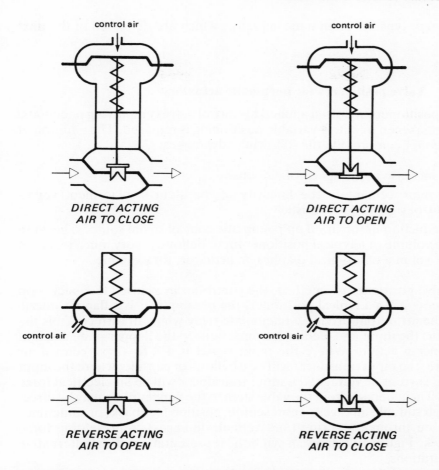

control air

control air

**DIRECT ACTING
AIR TO CLOSE**

**DIRECT ACTING
AIR TO OPEN**

control air

control air

**REVERSE ACTING
AIR TO OPEN**

**REVERSE ACTING
AIR TO CLOSE**

Fig 10.7   Four alternative types of action for actuators for flow control valves

electrohydraulic servovalve. Typically, the valve and actuator will be part of a position servosystem, with valve stem position feedback provided by a mechanically coupled rectilinear servo potentiometer or LVDT feeding an electronic amplifier which in turn drives the servovalve. Very much higher operating forces and speeds of response can be generated.

Electric motors are sometimes used as actuators, driving the valve stem through a suitable gear reduction. A position servo feedback loop may be added around this, using a servo potentiometer or LVDT and an electronic preamplifier and power amplifier. The speed of response will be relatively low.

Note that position feedback loops are applied only around valve actuators when infinitely variable positioning of the valve is required. Such a loop may be referred to as a 'positioner', although this term tends to be applied mainly to position (and force) feedback loops associated with diaphragm types of pneumatic actuator, which are discussed in the next section.

10.1.4   Valve positioners for pneumatic actuators

Valve positioners are often applied to control valves employing pneumatic actuators, when infinitely variable positioning is required. The addition of a positioner can provide the following advantages:

- an increase in the speed of response
- an improvement in the linearity of the stem displacement versus control signal characteristic
- a reduction in the drain on pneumatic control signal sources, because the volume of a typical positioner input bellows is very much less than the volume of a typical diaphragm actuator, for example.

The valve positioner is, in effect, the controller in a closed feedback loop arrangement. In a typical positioner, the preamplifier is a flapper-nozzle type, the air output driving a pneumatic relay which in turn controls the actuator: the input and feedback connections to the flapper plate result in differencing action. Where the input signal is a 3 to 15 psi control air pressure, the input transducer will be a bellows or capsule; where the input is a DC current or voltage, the input transducer will be an electrical force motor. The feedback from the valve stem to the flapper plate may be direct (through suitable linkages), representing position feedback, or sometimes it may be through a spring (and suitable linkages), representing force feedback. Figure 10.8 shows a symbolic representation of the alternative configurations.

   With any of these types, the feedback may be converted from linear to nonlinear by the insertion of a suitably contoured cam and follower into the coupling to the valve stem: the resulting nonlinear closed loop static characteristic will then supplement the installed characteristic of the control valve.

10.2   Pneumatic process controllers

Section 11.3 will give a detailed account of the configuration of general-purpose process controllers, treated in general terms. The specific implementation of these controllers, using pneumatic hardware, will be covered at this stage, in order to complete the exposition on pneumatic devices.

   A typical application of a pneumatic process controller is shown in Fig. 10.9. The function of the controller is to open and close the control valve so as to manipulate the inflow rate, in the presence of fluctuations in outflow rate. It does this in order that the liquid level in the tank, as measured by the transducer, shall match as closely as possible the desired

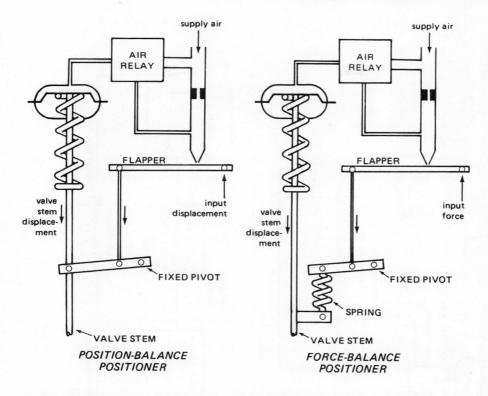

Fig 10.8 Symbolic representation of typical position-balance and force-balance control valve positioners

value, as determined by the manually adjusted set point. The functions of the pneumatic controller are:

- to enable the set point signal to be generated
- to receive the feedback signal representing the measured level
- to generate an error signal by comparing the above two signals
- to amplify the error signal and to incorporate dynamic terms, in generating the controller output signal.

The control law can incorporate one or more of the terms known as *proportional action*, *integral action*, and *derivative action*, which are described in Chapter 11.

Figure 10.10 is a symbolic representation of a controller containing only proportional action. In practice, the PV and set point bellows may be coupled (differentially) to the flapper through fairly complex linkage arrangements. The flapper-nozzle amplifier is the pneumatic equivalent to the electronic operational amplifier and its output pressure responds,

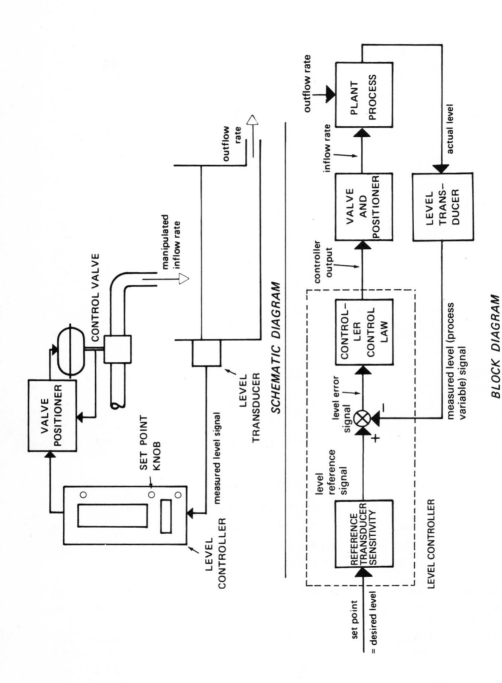

SCHEMATIC DIAGRAM

BLOCK DIAGRAM

Fig 10.9 Schematic and block diagrams for closed loop control of liquid level in a vessel, using a pneumatic process controller

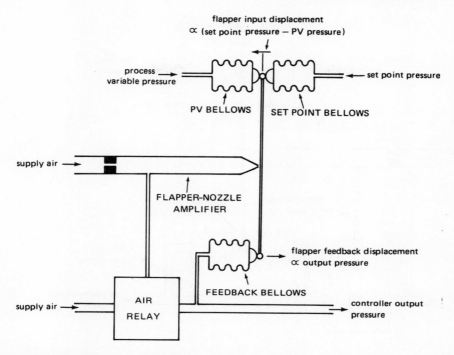

Fig 10.10 Symbolic representation of a pneumatic process controller incorporating only proportional action

nonlinearly, to minute changes in flapper displacement. The air relay behaves as a unity follower, so that its output pressure tracks the amplifier output pressure, but with significant increase in volumetric flow capacity. The feedback bellows completes a high gain negative feedback loop, and equilibrium is established by a force balance at the flapper. Thus, the controller output pressure is proportional to the difference between the set point and process variable pressures. The constant of proportionality may be adjusted by manually changing the moment arm ratios of linkages (not shown) which couple the feedback bellows to the flapper.

Integral action may be incorporated by adding a series connected combination of variable restriction and (integral action) bellows in the feedback path, as shown in Fig. 10.11. The restriction is analogous to a variable resistor and the bellows is analogous to a capacitor, so that adjustment of the restriction will cause the integral action time constant to be 'tuned'. If a second, variable (derivative) restriction is added at point X on Fig. 10.11, in series with the proportional action bellows, adjustment of this restriction will cause the derivative action time constant to be tuned.

Figure 10.12 shows the faceplate of a typical pneumatic indicating process controller, and the features shown are common to all general-purpose analog process controllers of the type described in Section 11.3.

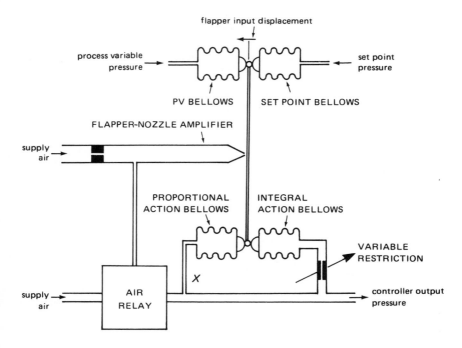

Fig 10.11　Symbolic representation of a pneumatic process controller incorporating proportional and integral actions

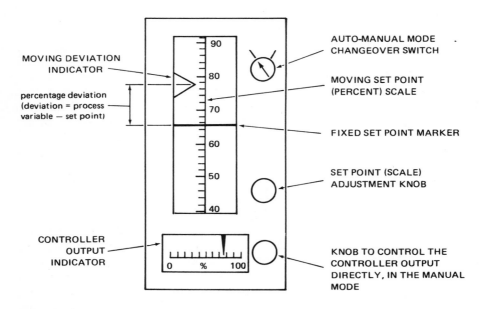

Fig 10.12　Faceplate of a typical pneumatic process controller showing instrument displays and manual controls

11

Electronic and Electrical Controllers

11.1 Classification

This chapter will deal specifically with electronic and other electrical controllers. Some of the principles to be described will also be applicable to certain equivalent pneumatic controllers, but detailed descriptions of these will not be included at this point: see Section 10.2 for details of pneumatic process controllers.

Commercially available general-purpose controllers can be subdivided into four categories, which will be considered in detail. Any other types of controller not described here are unlikely to be available off the shelf and would need to be custom designed and constructed.

The four principal categories of controller are *on-off temperature controllers*, *general-purpose process controllers*, *motor speed controllers*, and *sequence controllers*. These controllers may include provision for indicating the various signals involved (*indicating controllers*) and/or provision for chart-recording specific signals (*recording controllers*). Indicating instruments will usually have either a circular arc type of scale or, when a mechanical movement is involved, a 'thermometer' scale; however, in many modern installations the instrument will have been replaced physically by the simulation of an instrument scale on a video display, with the hardware including provision, using a keyboard, for setting the controller operating mode and parameter values and for selecting the variable to be displayed. Recording instruments are covered in detail in Chapter 7.

11.2 On-off temperature controllers

11.2.1 Temperature sensors

The simplest form of temperature sensor is the bimetal strip, described in Section 5.3.1. The strip may be either flat or a coiled flexure, with the change in the shape of the strip, resulting from changes in temperature, typically causing electrical contacts to be either closed or broken as required. Mechanical adjustments can be used to cause a change in the pre-stressing or pre-positioning of the strip, resulting in a change in the temperature at which the contacts just close (or open). Alternatively, motion of the strip can be used to reposition the moving member of a displacement transducer.

Another class of device involves using the expansion of a fluid in a container (for example, a Bourdon tube), resulting from applied heat, to activate either electrical contacts or a displacement transducer.

Thermocouples may also be used, with the EMF produced being used either to move a coil in a magnetic field or to provide the input signal of an amplifier. In similar fashion, resistance thermometers and thermistors may be used in a Wheatstone bridge circuit, with the unbalance voltage (resulting from temperature change) being used to move a coil or drive an amplifier.

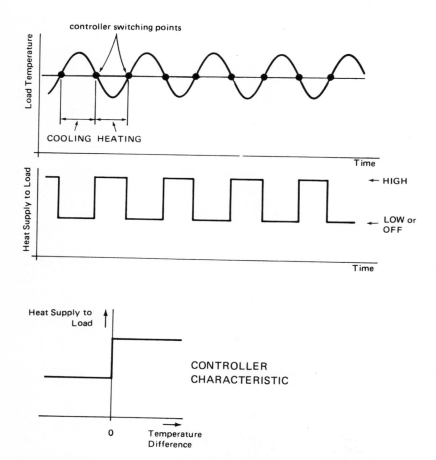

Fig 11.1   Cycling of load temperature and heat supply using a temperature controller having a simple on-off characteristic

11.2.2   Control laws

The simplest law involves switching on and off a heat supply, with switching in both directions occurring at the same (preset) temperature. This results in a cycling of the load temperature, as shown by Fig. 11.1. The magnitude and frequency of the temperature cycling will depend upon the magnitude of the heat source and the dynamics of the thermal load: in many installations, the thermal load will be such that the cooling interval is much longer than the heating interval, especially where cooling occurs entirely as a result of heat transmission to atmosphere.

A greater degree of control over the temperature cycling can be effected by designing the control law such that the switching point during heating occurs at a higher temperature compared with the switching point during cooling, as shown in Fig. 11.2. This type of law property is sometimes

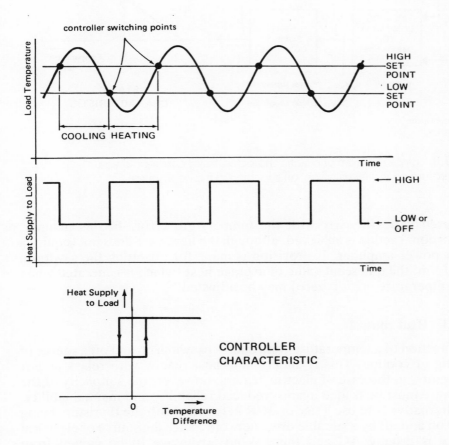

Fig 11.2 Cycling of load temperature and heat supply using a temperature controller having an on-off characteristic overlap (hysteresis)

referred to as 'overlap' or 'hysteresis'. Increasing the overlap zone will result in a reduction in the frequency of the temperature cycling, but this will be accompanied by an increase in the temperature excursion.

With 'multistep' control, the hardware makes provision for three or more switching temperatures, resulting in an equal number of different applied heat levels. Such a controller inevitably is more complex but can result in a significant reduction in the temperature excursion.

Proportional control may be effected, using on-off switching hardware, with the type of arrangement shown in Fig. 11.3. The mark-space ratio of the controlled heat supply changes as a function of the value of the temperature error voltage V_e, such that the average value of the heat

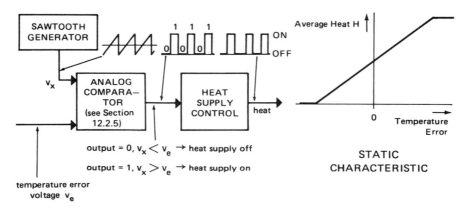

Fig 11.3　Block diagram and static characteristic for a proportioning temperature controller using on-off switching of the heat source

delivered varies linearly with the temperature error. In this manner, proportional action is achieved, although the hardware does not require a linear power amplifier. If provision is made for offsetting the sawtooth waveform, the quiescent value of average heat (which is generated when the temperature error is zero) may be adjusted.

11.2.3　Heat control

The function of a temperature controller is to switch on and off a source of heating or cooling. This is undertaken most readily with relays and/or contactors, in the case of electric heaters; however, the simplicity of the hardware must be traded against reduced reliability and maintainability. An alternative is to use Triac or SCR networks, with the thyristors being gated on and off by a suitable drive network: since the source of electrical energy is usually AC and there is no advantage to be gained from rectification, it follows that Triacs would be preferable to SCRs. Reliability and maintainability will be high with power semiconductors, provided that they are protected against excessive transients: the fact that a heater is almost purely resistive assists, in this respect.

Where the controller is concerned with the switching of steam or refrigerant flow, this can be effected by using a solenoid-actuated pneumatic pilot valve switching air to the diaphragm actuator of a control valve in the flow line.

11.3   General-purpose process controllers

11.3.1   History

General-purpose process controllers originated in pneumatic form about forty-five years ago. Subsequently, these were superseded by electronic versions, to a large extent, using firstly discrete component and then integrated circuit analog technology. Because of cost and reliability, recent development has involved synthesising the controllers with digital hardware, with the control law either hardwired or represented by program statements in software. The progressively increasing degree of complexity and sophistication has presented potential problems, in terms of maintenance, fault location, and repair, and these become significant factors when selecting a controller.

11.3.2   Types of control action

General-purpose process controllers can be subdivided into *feedback*, *cascade*, *feedforward*, and *ratio* controllers, depending upon the type of action which each is required to provide. In some cases, the type of action may need to be specified at the time of ordering, whereas, in other cases, it may be possible to configure one particular controller for any of these roles, at the time of commissioning. See Section 13.7 for descriptions of typical applications for the various controller types.

The application of a feedback controller is shown in Fig. 11.4. Such a controller makes provision for setting the desired value (called the 'set point'), accepting the measured value from the feedback transducer, determining the difference (called the 'error' or 'deviation'), and thereby generating an output signal which is related to the deviation by a preset algorithm. This algorithm usually includes time-dependent terms, thereby providing dynamic compensation of the control loop. This type of controller is providing feedback control in the conventional sense.

The application of a cascade controller, in conjunction with a feedback controller, is shown in Fig. 11.5. It will be seen that the two controllers must be similar, with the exception that the output signal from the feedback controller becomes the (variable) set point of the cascade controller, so that this set point would normally not be adjustable manually: it is said to be a 'remote' set point, as opposed to a 'local set point', which is set manually. The configuration represents a dual loop system, with the outer (major) loop having greater authority than the inner

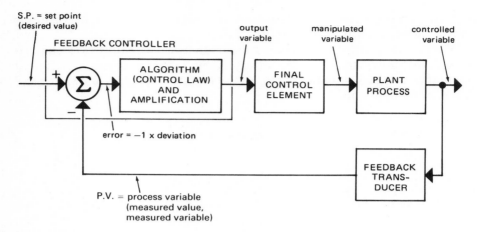

Fig 11.4   Block diagram of a feedback loop using a feedback controller

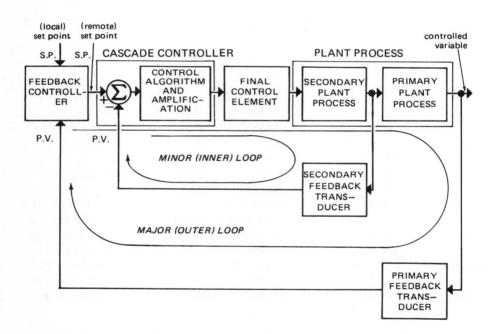

Fig 11.5   Block diagram of a two-feedback loop system using a feedback controller and a cascade controller

(minor) loop. Typically, the primary plant process would include a mathematical integration, so that the plant variable being measured by the secondary feedback transducer will then be zero whenever the controlled variable is stationary in value: the outer loop will dominate steady state control accuracy, whilst both feedback loops will determine transient behaviour. One or more of the following factors will be relevant, provided that the gain of the inner loop is made sufficiently high (see Section 8.5).

- The closed inner loop will have a faster transient response than that of the secondary section of the process, on its own, with the result that the speed of response of the outer loop will be enhanced.

- The closed inner loop will be relatively insensitive to changes in the properties of the secondary section of the process, so that this loop becomes an element having stationary properties, as far as the outer loop is concerned.

- The closed inner loop will be relatively linear in its behaviour, despite possible departure from linearity in the characteristics of the secondary section of the process, with the result that the linearity of the outer loop will be enhanced.

- Parasitic disturbances occurring within the secondary section of the plant will have a minimal effect on the output of the primary section of the plant process.

The choice of control algorithm for cascade controllers is usually made the same as that for feedback controllers, as discussed in Section 11.3.4.

A feedforward controller is used in order to anticipate, and correct for, large parasitic disturbances in a plant process, for which a feedback controller, on its own, proves to be inadequate. The output from the feedforward controller is used to augment the output from the feedback controller: typically, the output from the feedback controller is used as a remote input for the feedforward controller, as shown in Fig. 11.6. In the absence of the feedback controller, there ceases to be any feedback action, because there is no longer a closed loop. Ideally, the feedforward controller should be tuned so that the component of output signal which it generates from the measurement of the disturbance precisely cancels the direct effect of the disturbance upon the plant process. In practice, only an approximate cancellation is necessary, because the feedback action of the closed loop completed by the feedback controller should cope with any residual effect of the disturbance. The usual control algorithm provided by a feedforward controller takes the form of a 'lead-lag' law, corresponding to a transfer function of the type $K(1 + sT_1)/(1 + sT_2)$.

A ratio controller is used for controlling a process involving the mixing of two components A and B. The set point of the controller represents the desired value for the ratio A/B (say). Figure 11.7 represents two alternative configurations which theoretically could be used to achieve this end; however, the first configuration is not practicable, because both of the

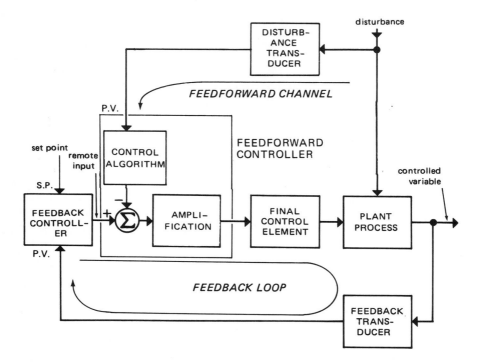

Fig 11.6   Block diagram of a feedback loop to which a feedforward channel has been added, to assist the loop with coping with large parasitic plant disturbances

transducer signals form part of the loop, with the result that the loop gain varies with signal level and is therefore far from constant in value. In the second system, the B transducer signal is external to the feedback loop, which therefore has a much more invariant loop gain: in this practical configuration, B is known as a 'wild variable', because it is assumed that it is this variable which is not being affected by the action of the final control element. The choice of control algorithm for ratio controllers is usually made the same as that for feedback controllers, as discussed in Section 11.3.4.

11.3.3   Signal levels

With electronic controllers, the range 4 to 20 mA is by far the most common for the process signals. The offset datum to some extent is inherited from the need for an offset datum with pneumatic signals but it does have the merit' of preserving a high signal/noise ratio and also provides means for detecting a broken connection. However, the control algorithms are usually processing voltage signals, referenced to signal common, rather than current signals, so that internal I/V and V/I conver-

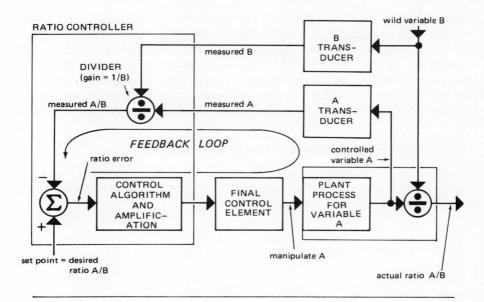

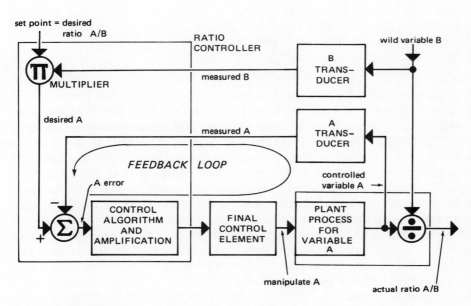

Fig 11.7   Block diagrams of systems using (above) an impractical ratio
controller configuration and (below) a practical ratio controller configuration

sion, together with datum offsetting, is normally incorporated into any controller designed for DC current transmission. For this reason, there will almost certainly be a steady trend towards DC voltages for transmission signals, especially now that more controllers are becoming digital: there is little sense in performing firstly an I/V conversion and then an A/D conversion at the input side, if one conversion can be eliminated by changing to voltage signals for transmission; a similar argument applies to the output side. Alternative, less common, transmission signal ranges include 0 to 20 mA, 10 to 50 mA, 0 to 10 V, 0 to 1 V, and 1 to 5 V DC.

11.3.4   PID feedback controller configuration

Figure 11.8 shows schematically the internal organisation of a typical analog *three-term feedback controller*. The algorithm is said to be 'three term' because it takes the typical form

$$\text{controller output} = \frac{1}{K_p}\left[e + \frac{1}{T_I}\int e\,dt + T_D\frac{de}{dt}\right]$$

where

controller steady state gain $= 1/K_p$
controller 'proportional band' $= 100K_p\%$
$T_I = $ 'integral action time', or 'reset time'
$T_D = $ 'derivative action time', or 'pre-act time'
$e = $ system error $= -1 \times$ 'deviation'

The proportional term e/K_p results in a component of output proportional to error; the integral action term $[1/K_p \cdot T_I \int e\,dt]$ results in a ramping component of output if the error is constant, and the derivative action term $[T_D/K_p \cdot de/dt]$ results in a steady component of output if the error is ramping. This last term tends to amplify any parasitic noise components which may be present in the signal representing error e, and therefore should be used with great caution whenever signals from feedback transducers are noisy: this is because noise may exhibit high values for instantaneous rates of change even when the peak value of the noise signal may be quite small. The derivative action term $[T_D/K_p \cdot de/dt]$ will generate an impulsive component of output, should the set point be changed suddenly, and, for this reason, some manufacturers prefer to connect the derivative action in the feedback path upstream of the error generation point, so that the set point component of the error signal can no longer be differentiated. This is shown in Fig. 11.9.

If the coefficients K_p, T_I and T_D can be adjusted independently of each other, each by means of its own potentiometer (say), the control law is said to be 'non-interacting': most modern electronic controllers are like this. In older electronic controllers, and in all pneumatic controllers, one control adjustment will change one coefficient in a major way but will also affect

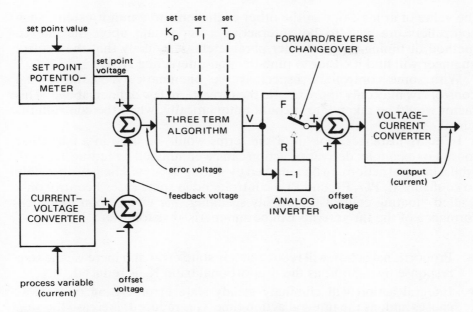

Fig 11.8   Block diagram showing the internal organisation of a three-term feedback process controller

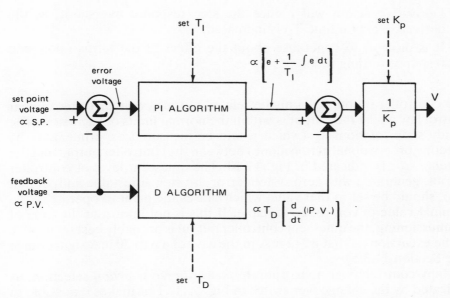

Fig 11.9   Reorientation of the derivative action into the feedback path of the controller, to prevent differentiation of sudden changes in set point

the value of at least one of the other coefficients to a minor extent: such controllers are said to be 'interacting'. Depending upon the chosen method of tuning the controller parameters, it is likely that the control engineer will find it easier to tune the non-interacting type.

With some controllers, especially the pneumatic types, the actual control law may only approximate the three-term law defined above. This factor would be taken into account automatically, when the controller is tuned.

In many instances, not all of the terms would be used in a particular loop. For example, derivative action can be eliminated by setting $T_D = 0$, whilst integral action can be removed by setting $T_I = \infty$. Thus, it is possible to configure P, PD, PI, and PID control, and in some cases I control only (called 'floating control'). Roughly speaking, the effect upon loop performance of the three terms can be summarised as follows.

- Proportional action will reduce steady state error and increase the step response overshoot, as the proportional band K_p is reduced.
- Integral action will eliminate steady state error arising from most causes and, as the integral action time T_I is reduced, increase the step response overshoot; it is unlikely to be required whenever the plant process contains an inherent integration term, since the resulting loop would then contain two integrations.
- Derivative action will reduce the step response overshoot, as the derivative action time T_D is increased.
- It is difficult to generalise about the effect of the terms upon step response settling time.

The controller output will only be related to the system error by a linear control law whilst the output is within the normal limits of excursion. The steady state relationship (which would be measured experimentally by selecting proportional action alone) between the controller output and the system error is indicated in Fig. 11.10. The quiescent level of controller output, generated with zero system error, is often adjustable and, in that case, should be set to that value which causes the plant to operate at the nominal value of controlled variable. If this is not known at the time of commissioning, the quiescent controller output is probably best set to 50% of the excursion — that is, 12 mA in the case of a 4 to 20 mA signal range (see Section 11.3.5).

Many controllers also incorporate *forward/reverse action* selection, as indicated by the changeover switch in Fig. 5.11. This makes it possible to include or omit a sign inversion, which can be used as necessary to ensure that the total number of sign inversions around the closed loop is an odd number, thus achieving negative feedback rather than positive feedback. Operation of the switch will have the effect of reversing, about the controller output axis, the characteristic of Fig. 11.10.

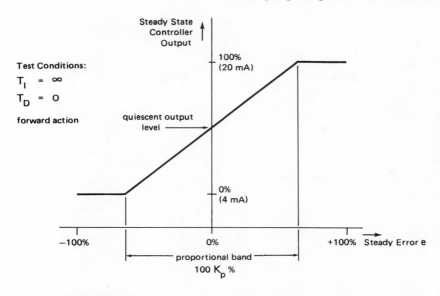

Fig 11.10   Steady state characteristic of a process controller showing the definition of proportional band and the quiescent output level

11.3.5   Auto/manual transfer

Often, when tuning a loop or when starting up or shutting down a plant, it is convenient to operate a system in an open loop mode (that is, with the feedback path rendered ineffective) rather than a closed loop mode. This is achieved by means of *auto/manual transfer*: in the manual mode, a special output offset adjustment is used to manipulate the value of the controller output signal and the three-term control law is by-passed (as shown in Fig. 11.11), thereby eliminating the feedback action. Typically, the same output offset potentiometer will be used to adjust the quiescent output level when the auto mode is selected: this can then be interpreted as a specific value for the constant of integration for the integral action term, when I action is being employed. Problems can arise when switching between the manual and auto modes, if this results in a discontinuity in the value of the controller output. Modern electronic controllers incorporate 'bumpless transfer' (or 'bumpless changeover') which, by electronic means, eliminates the possibility of a discontinuity in the output signal. Earlier controllers required manual 'balancing', whereby the new value of the output had to be adjusted manually before switching to the new mode, in such a way that no change in output value would occur at changeover. On very early controllers, the step in output value was unavoidable.

With some older controllers, it was possible for the hardware mechanising the integral action term to accumulate a non-zero value, when the controller was in the manual mode. This has been called 'reset wind-up', and can prevent a bumpless transfer. In a modern controller, the output of

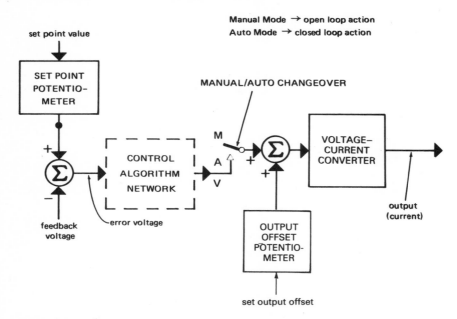

Fig 11.11   Modification of the block diagram of Fig 11.8 to incorporate auto /manual transfer and manual adjustment of output offset

the integral action term will be clamped at zero whenever the controller is in the manual mode. Note that the term 'reset windup' may also be applied to the integration of sustained values of error variable which might occur in the auto mode (when the loop is closed). This situation can arise when either large changes in set point or large changes in plant load are applied. The controller may include components to limit the value of the output signal generated by the network mechanising the integral action term.

11.3.6   Analog controller displays

Wherever displays are incorporated into the faceplate of an analog process controller, these are usually provided by moving coil instrument movements travelling over a strip scale, calibrated linearly in terms of percentages of full signal excursion. Figure 10.12 showed a typical layout of the faceplate of a pneumatic process controller, and a similar arrangement would be typical for an analog electronic counterpart. Some modern controllers employ solid state 'thermometer' types of scale, indicating 1% or ½% increments, using sets of parallel light-emitting diodes or miniature gas discharge indicators, each taking the form of a very short bar to represent a specific signal level. The variables usually displayed are set point, deviation, process variable, and controller output. Controllers often include means to indicate when the deviation exceeds manually preset upper and lower limits and, in addition, simultaneous output signals may be generated in order to activate remote alarm annunciators: alternatively, the alarm indication may be monitoring either the process variable or the controller output.

11.3.7 Digital process controllers

The availability of digital hardware has introduced improved reliability and flexibility to the production of general purpose-process controllers.

Where the number of control loops for a particular plant is small, it is becoming common practice to instal, in the place of an analog controller, an equivalent digital controller. With this approach, the digital controller would be designed to synthesise its analog counterpart and would resemble it physically, in terms of packaging, external connection, and commissioning procedures. Such controllers are microprocessor based, with a separate microprocessor being dedicated to each controller.

Where the number of control loops for an installation exceeds a figure in the vicinity of (say) eight, a more efficient approach is to use an electronic system in which the microprocessor is shared between a number of controllers, and this technique is often used in 'distributed control'. This approach offers much greater flexibility, because the controllers need not be configured until the commissioning stage, provided that a correct estimate of the quantity of controllers required has been made beforehand. Moreover, the configuration can be revised easily at a later stage, should this be desired. Configuring is undertaken using a hand-held or panel-mounted keyboard allocated to these controllers, or by means of a centralised keyboard console which would normally be sited in a control room.

The creation of a distributed control system is undertaken in stages: taken in chronological sequence, these would involve controller configuration and intercommunication, followed by the specification of ranges and alarm limits of variables, the values of the control law parameters, and the values of the set points. All such data would be entered manually, in the form of an alphanumeric code, and stored in protected semiconductor memory. Typically, communication between controllers and plant would be by means of analog signal transmission; communication between controllers situated within the same logical grouping would be by means of parallel digital data transmission; communication between logical groups of controllers, central control stations, and process computers would be by means of serial digital data transmission: see Section 2.10 for discussion on digital data transmission techniques.

In addition to synthesising all of the functions available with analog process controllers (Sections 11.3.2 to 11.3.6 inclusive), the distributed control configurations typically offer the following additional types of function, all of which would be implemented digitally:

- nonlinear static characteristics, involving such nonlinearities as deadspace, square laws, etc.

- summation, multiplication, division, square-rooting, etc. of data values
- gain constants and time constants which can be made variable functions of specified input signals
- *supervisory control*, in which the (remote) set point values can be set electronically, using data generated by an on-line process computer
- *direct digital control* (DDC), in which the operation of the microprocessor-based controller is now taken over by an on-line process computer, with the controller relegated to providing back-up capability in the event of computer failure.

The ability to introduce nonlinearities into controller static characteristics can effect the following types of improvement:

- control loops can be made less sensitive to parasitic noise
- the settling times of control loops, following set point or load disturbances, can be minimised
- loops can be made less sensitive to load disturbances
- the dynamic behaviour of loops can be made less sensitive to control law parameter settings, set point values, and load values
- compensation can be made for the effects of nonlinearities occurring in the final control element, plant process, or feedback transducer
- the effect of interaction with other control loops can be minimised.

The ability to make gain constants and time constants signal dependent enables a simple form of adaptive control to be introduced, whereby the form of the control law can be modified to accommodate sensed changes in plant operating conditions. This capability also facilitates improved start-up and shut-down procedures, whereby the controller settings ideally should be 'scheduled' to follow a predetermined sequence: typically, this scheduling would be supervised by a process computer or, in some cases, a *programmable logic controller* (see Section 11.5.3).

More recently, a range of self-tuning microprocessor-based process controllers has been released onto the market. These employ routines which involve the automatic updating of the PID parameters following either on-line analysis of the dynamic properties of the plant process or comparison of the current behaviour of the control loop with that of an idealised mathematical model specified at the time of commissioning. In some cases, parameter updating may not proceed automatically until the operator has had the opportunity to indicate (using a key) that the new set of parameter values proposed is indeed acceptable. Some of these controllers generate electronically test disturbances of specific waveform to be injected, at the controller output, into the loop in order to identify the current status of the loop: clearly, such a disturbance has the potential to upset momentarily the quality of the product of the process.

Because of the increased flexibility afforded by digital process controllers, the following types of decision may need to be included, when the hardware is being selected.

- In the event of a failure in a supervisory computer, should the controller stay in the auto mode, holding the last value of the set point?
- In the event of a failure in the controller, should the output line hold the last value of the controller output?
- Should each controller be self-contained or can controller hardware be distributed and shared between various printed circuit assemblies?
- Should each controller have its own independent power supply?
- If power supplies are shared, should there be back-up power supplies?
- What integrity should data highways have, in terms of protection against damage and susceptibility to electrical noise?
- Should controllers be separate physically from the supervisory computer, or be an integral part of the computer interface?
- Should controllers be self-monitoring, and should provision be made for their duplication, for failure-survival purposes?
- Should controllers have their own manual controls and displays, or can these be integrated into a centralised control console?
- Should controller internal configurations be modular, to facilitate rapid fault isolation and rapid return to service, using possibly low grade personnel?

It can be seen that the introduction of digital controllers can involve a number of types of decision which are not relevant to the selection of their analog counterparts.

11.4 Speed controllers

11.4.1 General-purpose speed controllers for DC motors

Figure 11.12 is a generalised representation of the various alternative arrangements used in DC motor speed control systems. The normal practice with large motors is to configure for armature control, with the armature power being derived from the AC mains and rectified to DC using a suitable *converter*. The type of converter to be used will depend to some extent upon the power rating of the motor: converters employing either bipolar power transistors or power MOSFETs can now be built to control motors having ratings up to many tens of kilowatts; SCR converters have been used extensively for all motor sizes up to several megawatts, although they are less likely to be used in the future at the lower end of the power range, due to the increasing competitiveness of transistor and

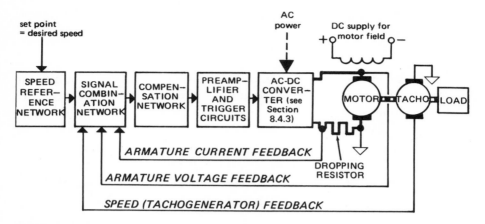

Fig 11.12   Schematic diagram of a generalised DC motor speed control system incorporating alternative minor feedback paths

MOSFET alternatives. The design of the converter will take into account the following factors:

- the motor rating
- the motor duty cycle
- the possible need for reversibility of the direction of rotation
- the type of braking action required
- the waveform of the AC line current, which will have a 'chopped' appearance — this waveform may be made to approach the desirable sinusoid by progressively increasing the pulse number created for the converter supply
- the possible need for electrical isolation between the AC mains and the motor circuit, so that the latter may be earthed at that point most convenient from controller design or safety aspects.

When the motor is operated under conditions of constant excitation, a speed range in the region of 20:1 is generally regarded as being a maximum. The lower speed limit is determined by the slow running performance of the motor, which is subject to 'cogging': this means that the motion is jerky, due to the effect of the armature slots on the time-variation in the magnetic flux distribution. The speed range may be extended to (say) 100:1, relative to the base speed, by weakening the field as a function of the set point (the desired speed), using the technique indicated in Fig. 11.13.

The feedback paths shown in Fig. 11.12 have the following properties.

- The primary (negative) speed feedback is obligatory and is normally derived from a DC tachogenerator. If reduced accuracy can be toler-

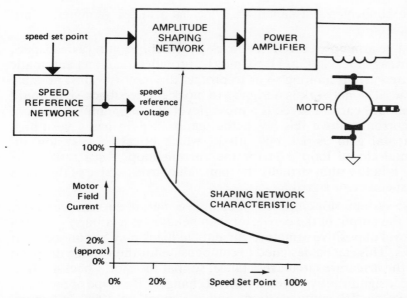

Fig 11.13 Mechanisation of field weakening for a DC motor in order to extend the available speed range

ated, the motor back-EMF can be measured, using the bridge arrangement described in Section 3.4.6, and the derived voltage used as the feedback signal: this technique is not appropriate when field weakening is incorporated, since the EMF is then no longer a function of speed alone.

- Negative armature current feedback can be used to change the converter into a current source, as opposed to a voltage source: this can result in improved motor dynamic performance. (See Section 8.5.3 for a detailed discussion). One disadvantage with the voltage source is that, if the armature supply voltage is subjected to a sudden change (arising from a sudden change in speed error), this can generate potentially damaging current levels circulating through the motor armature. This situation occurs whenever the motor back-EMF is significantly different in value from the applied armature voltage. If the gain around the current feedback loop is sufficiently high, it will have a linearising effect upon the converter which, on its own, can possess a nonlinear relationship such as would result, for example, from the law associated with the phase control of SCRs, as is discussed in Section 8.4.3.

- Negative armature voltage feedback may be switched in, on standby (when the set point will have been switched to zero), to ensure that the armature voltage, and hence the motor speed, is held rigidly at zero. In some systems, negative armature current feedback may be omitted (so that the output from the converter will probably then behave as a voltage source), in which case negative armature voltage feedback may be connected permanently, to act as a minor (compensation) feedback path.

- A negative armature current feedback path incorporating a deadspace element, as shown in Fig. 11.14, may be provided, to effect automatic armature current limiting so as to protect the motor against possible overload. The network is designed to block the feedback signal until the armature current exceeds a preset level of, for example, 125% of rated current: above this level, the feedback path has a very high incremental gain associated with it, which effects a very low incremental closed loop gain for the current loop. As a result, the current is held within virtually the limit value, irrespective of the value of the speed error signal.

- In those systems not employing negative armature current feedback, so that the output of the converter then behaves as a voltage source, a small level of positive armature current feedback may be connected at all times. This can be designed to compensate for the voltage dropped across the armature circuit resistance, so that the motor back-EMF is no longer significantly affected by load changes. This type of feedback is known as 'IR compensation' and it cannot be 100% effective,

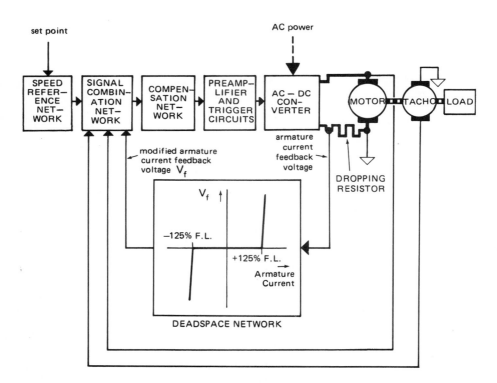

Fig 11.14 Mechanisation of armature current limiting feedback using a deadspace network

because of the additional (nonlinear) effect of armature reaction upon the motor back-EMF.

Other facilities which are often built into commercial speed controllers include the following.

- Selectable set points, facilitating sequence control of speed. See Section 3.3.1 for details.
- Dynamic limiting of set point changes, to limit demanded motor acceleration. This is achieved by inserting a low-pass R-C filter between the speed reference network and the speed error generation point.
- Adjustment of minimum and maximum speed set points. Again, see Section 3.3.1 for details.
- Adjustment of compensator parameters, equivalent to the tuning of general-purpose process controllers.
- Regenerative braking, which requires that the directional sense of the motor torque be controlled in a manner causing a return of energy to the AC supply, when this is required. Such an arrangement, whilst increasing the complexity of the converter, will improve the system response time to a demanded reduction in speed and/or a demanded reversal in the direction of motion. An alternative is dynamic braking, with which a high dissipation resistor is switched across the motor armature, in order to provide a sink for the energy stored in the motor and load, but this method is unacceptably wasteful, except for small motors.

11.4.2 General-purpose speed controllers for AC motors

In terms of capital cost, a DC motor is much more expensive than an AC motor of comparable rating; conversely, a speed controller for an AC motor has, until recently, been much more expensive than a comparable speed controller for a DC motor, because of the greater complexity associated with the former. In terms of total capital outlay, DC drives have had the financial edge over AC drives, although against this should be weighed the greater reliability and maintainability associated with AC motors. Recently, however, technological improvements and reduced device costs have made AC motor controllers much more competitive, especially when the combined cost of the controller and the motor is considered: when reliability and maintainability are also taken into account, the AC drive is becoming increasingly the more attractive of the alternatives, especially in the power range from 1 kW to several tens of kilowatts.

Conventional AC motors can be divided into four categories, all of which can be subjected to electronic speed control: induction motors, AC commutator motors, synchronous motors, and reluctance motors, the last named being a type of synchronous-induction motor. However, AC commutator motors are not competitive when employed with electronic speed control, so that they will not be considered further here.

With induction motors, the usual value of standstill inrush current can be limited with electronic control, by exciting the stator with reduced voltage or reduced frequency during starting. With electronic control, synchronous and reluctance motors need not be started in an induction machine mode, because of the variable frequency action inherent with electronic controllers for these types of motor.

Electronic controllers for conventional induction motors can be divided into three categories. In each case, the converter and inverter networks used may employ SCRs or, certainly at the low end of the power range, bipolar power transistor or power MOSFET networks may be used as alternatives. The three alternative control strategies are as follows.

Manipulation of stator voltage

Figure 11.15 shows a family of torque vs speed characteristics for varying levels of stator voltage, for a conventional induction motor. For a load exhibiting a constant torque, the speed range would be limited to the region between the operating points A and B, which would usually be unacceptably restrictive. However, with fan types of load, the speeds represented by the region between points O and C can cover a wide range, so that this type of strategy can be suitable for fan loads. Efficiency is low and internally generated heat is high at low speeds, because of the correspondingly high slip frequencies, so that this control technique is not appropriate for large motors. Figure 11.16 represents a speed control system using stator voltage manipulation: other (minor) feedback channels may also be present in a practical system, but these have been omitted

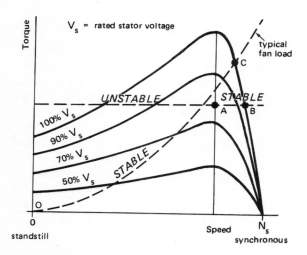

Fig 11.15 Family of torque vs speed characteristics for a conventional induction motor for varying levels of stator voltage

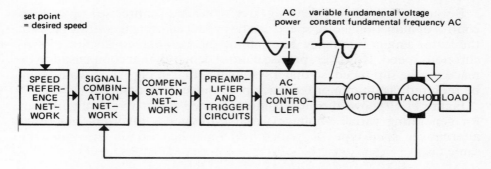

Fig 11.16 Schematic diagram of an induction motor speed control system involving manipulation of stator voltage

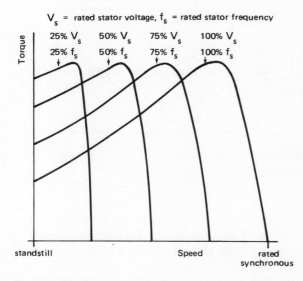

Fig 11.17 Family of torque vs speed characteristics for a conventional induction motor for varying levels of stator voltage and frequency

from the diagram. The AC line controller blocks part of each half cycle of the AC supply waveform, so that the fundamental component of the (constant frequency) motor voltage waveform is manipulated.

Manipulation of stator voltage and frequency

Figure 11.17 shows a family of torque vs speed characteristics for varying levels of stator voltage and frequency, with both being varied in the same proportion. With this strategy there is almost no restriction upon matching the motor to the load requirements, within the available torque and speed ranges, and the motor will be operated at low values of slip for all levels of excitation, yielding high efficiency and low heat generation.

Because of the low operating levels of slip, (imprecise) open loop control of induction motor speed can be produced using this strategy, with the stator supply frequency determining the instantaneous value of synchronous speed. However, precise control demands that compensation be made for the slip speed, so that a system like that in Fig. 11.18 would then be required: again, minor feedback paths have been omitted.

Figure 11.19 shows the most commonly used alternative configurations for generating variable-voltage variable-frequency inverters. As other alternatives, some types of *cycloconverter* may be used for performing the same task.

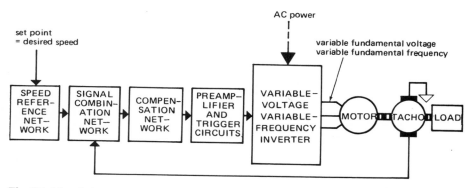

Fig 11.18 Schematic diagram of an induction motor speed control system involving simultaneous manipulation of stator voltage and frequency

Manipulation of rotor energy

This method is suitable for wound rotor induction motors, for which the transfer of energy from the rotor back into the AC mains is manipulated by means of a line commutated inverter, as shown in Fig. 11.20.

The effect upon the family of torque vs speed characteristics of transferring increasing levels of energy out of the rotor is comparable to that of increasing the rotor circuit resistance, which is demonstrated in Fig. 11.21. However, electrical energy is no longer dissipated as heat in the rotor but is transmitted back to the supply, so that the power efficiency is high despite the fact that high values of slip can occur during operation. In more complex arrangements, energy transfer between the mains and the rotor can be made bidirectional, yielding families of torque vs speed characteristics equivalent to both positive and negative values of rotor circuit resistance.

Synchronous motors and reluctance motors are controlled by manipulation of frequency but not voltage. For these motors, therefore, simplified versions of the controllers of Fig. 11.19 can be used, omitting the variable voltage facilities. In the absence of control over the full frequency range, a synchronous motor is not self-starting and then requires to be operated in an induction motor mode during starting: the necessary hardware can be

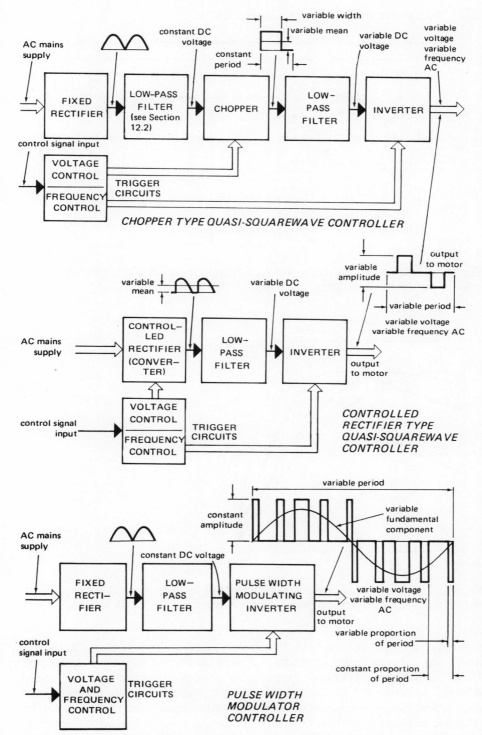

Fig 11.19 Three alternative configurations for generating simultaneous manipulation of stator voltage and frequency, for an induction motor speed control system

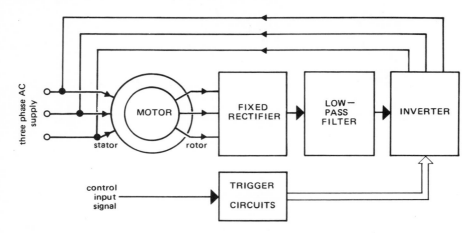

Fig 11.20   Schematic diagram of a system for manipulating rotor energy in a wound rotor induction motor

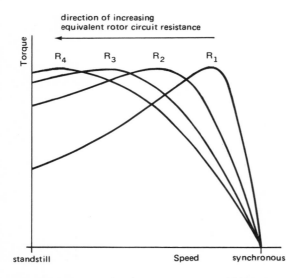

Fig 11.21   Family of torque vs speed characteristics for a wound rotor induction motor with manipulation of rotor energy

produced using an integral AC brushless exciter, a fixed rectifier network, and solid state switching circuits, all mounted upon the rotor assembly. With both types of motor, open loop action will yield precise speed control, because of the exact relationship between rotor speed and stator supply frequency, once synchronous operation has commenced. It may be possible to operate several motors from the one controller, with the rotors remaining in precise synchronism with each other, once synchronous action has started.

AC motors may be braked using one of the following alternative techniques.

• Alternately reversing the phase sequence of the stator connections, a procedure known as 'plugging'. This is unsuitable for synchronous motors.

• Disconnecting the normal AC supply from the stator and, instead, applying a DC voltage. This represents a form of dynamic braking, and is suitable for induction motors.

• Disconnecting the normal AC supply from the stator and, instead, connecting suitable resistors across the stator terminals. This represents a form of dynamic braking, and is suitable for synchronous motors.

• Disconnecting the normal AC supply from the stator and, instead, connecting the stator through a fixed rectifier network into a suitable DC source. This is a form of regenerative braking, and is suitable for synchronous motors.

• Gradually reducing the frequency of the stator supply. This is a form of regenerative braking, and is suitable for all three types of AC motor.

In all cases involving switching, the braking circuitry can be implemented using either electromechanical or solid state switches.

11.4.3   Incremental controllers for stepper motors

The control of a stepper motor (described in Section 8.6.3) involves the DC excitation of the stator windings, in a predetermined sequence. Each manufacturer of stepper motors has tended to adopt individual standards, so that the multiplicity of stepper motor types which has resulted means that there are no universal controllers for these motors: each motor tends to require its own controller design.

Certain properties of stepper motors are common.

• The speed of rotation will be proportional to the frequency of the pulse train applied to the drive circuits, provided that a specified maximum frequency and load torque are not exceeded.

• The position of the motor rotor will depend upon the number of pulses which have been applied previously to the drive circuits, provided that the motor has been able to develop sufficient torque to enable it to respond to every pulse.

• The direction of rotation can be reversed by reversing the sequence in which the motor coils are energised and de-energised.

• The motor coils are highly inductive, so that the drive transistors must be protected against voltage spikes whenever a coil is de-energised.

• By energising two adjacent motor coils simultaneously, the rotor step size may be halved in value. The consequent reduction in kinetic energy will result in less overshoot of each step position. The frequency of the pulse train will need to be doubled, in order to restore the original stepping rate.

• Motor efficiency can be improved by reversing the current in each motor coil, as an alternative to switching the current off. With such a bipolar drive (which requires greater circuit complexity), the total input power to the motor may be half that required for the corresponding unipolar drive, for approximately the same motor performance.

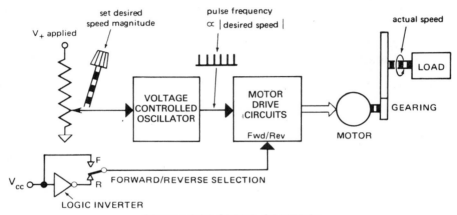

OPEN LOOP SPEED CONTROL

CLOSED LOOP SPEED CONTROL

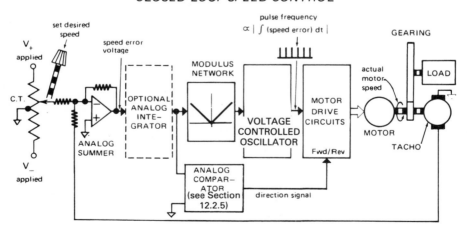

Fig 11.22 Schematic diagrams for alternative open loop and closed loop speed control regimes for a stepper motor

Figure 11.22 shows representative configurations for open loop and closed loop speed controllers. The motor drive circuits receive a pulse train having a frequency proportional to the voltage applied to the voltage-controlled oscillator. In the case of the open loop configuration, the *forward/reverse* switch would not be required for a unidirectional drive. More precise control of speed might be achieved with the closed loop configuration, which would almost certainly require the analog integrator in order to achieve satisfactory operation: in the steady state, this integration will develop that steady output voltage required by the oscillator in order that the motor may be stepped at precisely that speed corresponding to zero speed error. In Section 13.4.3, the application of stepper motors to position control systems is described, and further examples of speed control applications are given in Section 13.5.3.

The pulse train from the voltage-controlled oscillator would be applied to a purpose-designed ring counter, an example of which is indicated in Fig. 11.23. Typically, at any time one output would be high whilst all other

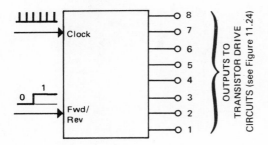

Fig 11.23 Use of a ring counter to produce a cycling sequence of signals for exciting transistor drive circuits for stepper motor stator windings

outputs would be low, although the reverse situation might be preferable for some designs. The input pulse train would cause the high signal to be stepped around the outputs in numerical sequence: 1, 2, 3, 4, 5, 6, 7, 8, 1, 2, etc. when the fwd/rev input is high (say), and 8, 7, 6, 5, 4, 3, 2, 1, 8, 7, etc. when the fwd/rev input is low. In other cases, two outputs might need to be high at any one time, with the others driven low, and these two high outputs would be circulated in sequence by the input pulse train. In further cases, the outputs might need to be high alternately singly and in pairs: many different combinations are possible. Typical pulse frequencies would be in the 300 to 400 Hz range for permanent magnet stepper motors and the 700 to 800 Hz range for variable reluctance stepper motors; however, stepping rates as high as 20 kHz have been reported, the upper limit on speed being dependent upon the time constant of the stator coil circuits.

The outputs from the ring counter are used to excite transistor drive circuits, there being one for each motor coil or for each pair of coils, typically. To limit the effect of the winding inductance, it is preferable that each winding circuit should present a switched current source to the coil. In the case of variable reluctance motors, some current must be retained when the motor is required to remain stationary, if a holding torque is required. In the case of permanent magnet motors, however, some holding torque is present without energisation, so that current pulses of controlled mark-space ratio can be used to excite the motor coils, with the resulting reduction in the mean level of current being used to minimise the

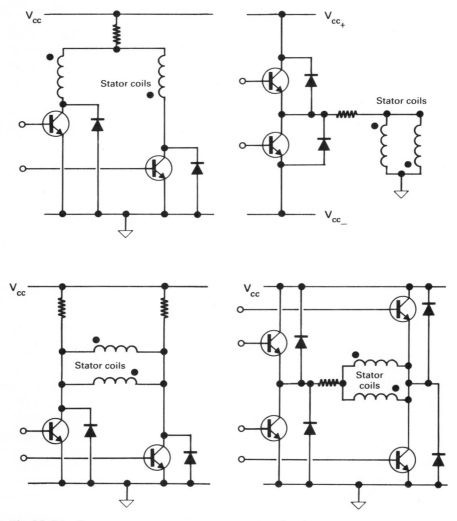

Fig 11.24   Four representative transistor circuits for driving stepper motor stator coils

heat developed by the motor. Figure 11.24 shows representative transistor drive circuits, some of which allow for two windings to be energised simultaneously; many other configurations are possible. Diodes, zener diodes, and resistors may be incorporated into the circuits, to provide protection for the transistors, to provide discharge paths for the coil inductances, to alter the circuit time constant, and to modify (that is, dampen) the typically oscillatory response of the motor rotor to a step in the position of the stator flux.

Any gearing and/or couplings in the drive between the motor and the load will be subjected to a hammering type of action, and allowance must be made for this. If the gearing is omitted, most applications would require the use of a motor designed to provide many steps per revolution: the large number of coils occurring in many such motors would then result in the need for an equally large number of transistor drive circuits.

Stepper motors are prone to overheating, if subjected to continuous high-speed operation, so that the duty cycle is an important facet of their use. The fact that significant energy can be wasted, in the form of heat dissipation, is an indication that the power efficiency of these devices can be low in value.

11.5 Sequence controllers

11.5.1 Motor-driven electromechanical timers

These represent the simplest form of sequencer. Usually, the motor runs at constant speed and turns a bank of rotary switches. The sequence is predetermined, to some extent, by the configuration of the switch wafers, but the sequence can be modified by changing the external wiring to the switch wafers. This type of sequencer has the merit of simplicity but life, reliability, and maintainability will be limited. This type also has the merit of making it easy to combine various types of voltage supply and various types of load, and easy to obtain isolation between the various supplies.

11.5.2 Relay and contactor networks

These networks normally employ straightforward 'ladder' arrangements of relay/contactor contacts and coils. Adjustable time delays can be a problem, especially if long periods of dwell are required. These networks have the advantages of ease of combination of types of supply and types of load, together with ease of electrical isolation of supplies. Life, reliability, and maintainability can be good, depending upon the type of environment and the switching duty required of the contacts. These three factors can be improved by using either dry reed relays or mercury wetted relays, at lower power levels, and improved still further by using solid state relays and contactors, which employ power MOSFETs, SCRs, and Triacs. In the case of solid state devices, problems may arise with signal isolation requirements (which can be alleviated by using opto-isolated components), with breakdown due to voltage spikes, and with false switching caused by crosstalk.

11.5.3   Programmable logic controllers (PLCs) and programmable controllers (PCs)

PLCs and PCs are constructed using dedicated microcomputers and are used for providing solid state sequence control. PLCs are intended to replace ladder networks of electromechanical relays and contactors. PCs have the same capabilities as PLCs but incorporate many additional features, including the capacity to process numerical data at bit, byte and word levels; some PCs include analog input/output channels and software to provide digital three-term process control.

Typical functions which a top-of-the-range PC might provide for processing numerical data stored in holding registers could include:

- ADD   SUBTRACT   MULTIPLY   DIVIDE
- DOUBLE PRECISION ADD   DOUBLE PRECISION SUBTRACT
- REGISTER–REGISTER MOVE   CLEAR REGISTER
- LEFT SHIFT   RIGHT SHIFT
- FIRST IN — FIRST OUT QUEUING
- logical arithmetic   matrix arithmetic
- searching for data values
- bit manipulation.

The range of PLC and PC products commercially available is considerable, and space permits further consideration only of one typical category of top-of-the-range PLC. This particular type consists of a *processor unit*, communicating through a parallel data bus with one or more *interface units*, and through serial communication channels with a *programming unit* and printers, which are used for generating records of ladder diagrams and for logging in-service events.

The programming unit is a dedicated microcomputer and is normally used only when the ladder network is being configured: for this reason, it can be time-shared between a number of processor units. It may be used to communicate with them using a telephone line and modems. The programming unit includes a monochrome screen and keyboard, and stores the program which will compile the ladder diagram data into object code, to be downloaded into, and stored in, the processor unit. A disc drive permits loading of the compiler and production of back-up storage of the ladder diagrams generated by the programmer. Programs may also be input in the form of Boolean equations, rather than as ladder diagrams.

The interface units are rugged assemblies installed close to the plant and have a backplane carrying the parallel bus of the processor unit, which is situated nearby. Using plugs and sockets, a set of interface modules is installed in the backplane. Typically, a module provides a set of input or output channels for one of the following types of signal:

- TTL-level DC input or output voltage
- low-level DC input or output voltage
- high-level DC input or output voltage
- low-level AC input or output voltage
- high-level AC input or output voltage.

Incoming wiring from on/off sensors, such as limit, pushbutton and toggle switches on the plant, is connected to terminals on the input modules. Outgoing wiring to on/off actuators and visual indicators on the plant is connected to terminals on the output modules. Power to excite all input and output circuits is supplied from external sources. Input and output modules have a high level of signal isolation and over-voltage and over-current protection designed into their circuitry.

The processor unit is a dedicated and ruggedised microcomputer which stores the compiled code in battery-backed memory. When the operator commands the code to execute, it does so by scanning through the ladder diagram in spatial sequence, rung by rung, updating the output states as a result of interrogating the latest states of input devices and intermediate elements in the program.

Input signal states are interpreted on the ladder diagrams as the states of contacts on imaginary relays. Output signal states are indicated as the states of imaginary relay coils. Alternative relay contact types available are *normally open*, *normally closed*, *transitionally off* (that is, closing for one scan time for a high–low transition of the activating signal), and *transitionally on* (that is, closing for one scan time for a low–high transition of the activating signal). Alternative relay coil types are *normal* and *latching*. Coils may have an unlimited number of associated contacts, the combination representing an imaginary relay. The total complement of elements on a ladder diagram is limited solely by the memory size of the unit. The diagram is constructed, on the CRT screen, in pages: communication between pages is achieved by having the coil of an imaginary relay on a preceding page and an associated contact on a succeeding page. Upcounters, downcounters and timers may be added to the ladder diagram.

Figure 11.25 shows a page of a typical ladder diagram. Power is considered to flow either from left to right or vertically, but not from right to left, from the power rail to an implicit signal common down the right-hand side of the diagram. The diagram is built up systematically, using cursor keys to position a symbol, numeric keys to allocate addresses within permitted ranges, and hard and soft keys to select a specific symbol, including horizontal and vertical shorting links. Any symbol except a relay coil can be associated with the first ten columns, numbering columns from

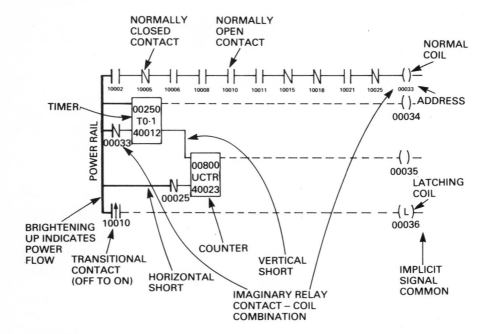

Fig 11.25 Example of a ladder diagram for a typical programmable logic controller

the left. Only coils may be placed in the eleventh column. Programming and execution commands are selected using a combination of hard and soft keys: the current status of the set of soft keys is displayed on the screen.

The program may be checked systematically during commissioning by intentionally using soft keys to force each coil, in turn, into an energised or de-energised state: the states of other elements on the diagram may be interrogated either by displaying them in tabular form or by noting the brightening up of interconnections and symbols on the diagram, signifying (imaginary) power flow. The states may also be displayed during normal operation, if the programming unit is left connected: otherwise, the unit may be removed once the ladder diagram has been checked out and placed in an executing state.

Factors to be taken into account when selecting a PLC include:

● input and output signal capacity
● the availability of suitable interface hardware
● the degree of difficulty involved with formulating programs
● the facilities available for entering and displaying programs and system status
● the ability to expand subsequently the system size

- the ability to modify the program easily
- ease of fault identification (including fault diagnostics) and isolation, and of return to service
- survival of power supply failure
- reliability and maintainability.

When correctly used, PLCs offer the advantages of long life, high speed of operation, compactness, ease of program reconfiguration, and minimal power requirement.

12

Hardware to Generate Sum and Difference Data; Mechanical Components

12.1 The combination of data

In any closed loop system, it becomes necessary to combine data. For example, it is always necessary to compare the controlled variable data (representing actual value) against the reference variable data (representing desired value). In systems containing multiple feedback paths, the minor feedback data must be added or subtracted at appropriate points in the forward path of the system, in the usual arrangement.

In many instances, it is more convenient to add, rather than subtract, the signals representing the data to be compared. The summation of data can be arranged to implement the subtraction of data by firstly inverting the sign of the data to be subtracted. This is demonstrated by the equivalence of the two diagrams shown in Fig. 12.1. The inversion required for data b

Fig 12.1 Subtraction of data using an addition process with sign inversion

would typically be implemented by the insertion of sign-inverting hardware or, alternatively, by modifying either the input connections or the output connections of the transducer generating b (assuming this to be feasible).

Often, it is not practicable to combine data in one-to-one proportions. This situation can be demonstrated by referring to Fig. 12.2. Suppose, for example, that the diagram represents the instrumentation of an analog speed control system, with the reference transducer being a potentiometer and the feedback transducer a tachogenerator. Therefore, both K_{tr} and K_{tf} will have units of volts per rad/s. Suppose, also, that these sensitivities are such that, when a full speed of 200 rad/s is both demanded and achieved,

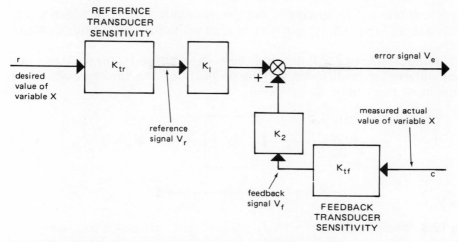

Fig 12.2 Combination of data in proportions other than one-to-one

$V_r = 10\ V$ and $V_f = 100\ V$. In this situation, the error $e(= r - c)$ is zero, so that the error signal V_e must also be zero. Now

$$V_e = K_{tr}K_1 r - K_{tf}K_2 c$$
$$= K_1 V_r - K_2 V_f$$
$$= 0$$

It follows that

$$10K_1 - 100K_2 = 0$$

so that

$$\frac{K_1}{K_2} = 10$$

Thus, the two signals should be combined in inverse proportion to the sensitivities of their transducer sources:

$$K_{tr} = \frac{10}{200}, \qquad K_{tf} = \frac{100}{200}, \qquad \frac{K_{tr}}{K_{tf}} = \frac{1}{10} = \frac{K_2}{K_1}$$

It should be noted that the calibration of the reference transducer relates to the scale arbitrarily marked in units of rad/s (or rpm), against which the operator manually sets his desired speed, by adjusting the shaft of the potentiometer. Normally, full scale deflection would represent maximum desired speed.

Note that the diagram of Fig. 12.2 can be rearranged mathematically to represent one-to-one mixing of data, as shown in Fig. 12.3, which represents a unity feedback system: that is, one with unity gain in the feedback path.

Having established that data may be both added and subtracted and that they will need to be mixed in varying proportions, the hardware which can implement this can be investigated.

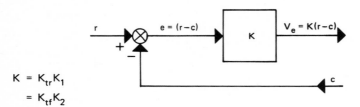

$$K = K_{tr}K_1$$
$$= K_{tf}K_2$$

Fig 12.3 Rearrangement of block diagram of Fig12.2 to create a unity feedback system

12.2 Electrical methods for combining analog signals

Analog electrical signals can be combined in terms of combining either voltages, currents or electro-magnetic fields. In modern technology, voltages and currents are usually combined using analog computing networks based on integrated circuit operational amplifiers. Some representative examples will be given.

12.2.1 Inverting summer configuration

Referring to Fig. 12.4 and assuming an 'ideal' operational amplifier, it is readily shown that

$$V_o = -\frac{R_f}{R_1}V_1 - \frac{R_f}{R_2}V_2 = -(K_1V_1 + K_2V_2)$$

This network is suitable for adding both DC and AC voltages. The loads on the input signal sources will be R_1 and R_2 ohms, respectively, provided that the amplifier output is not driven into saturation. Both input signal sources need to be single-ended and referenced to common.

12.2.2 Non-inverting summer configuration

Referring to Fig. 12.5, it is readily shown that

$$V_o = \frac{\left(\dfrac{V_1}{R_1} + \dfrac{V_2}{R_2}\right)\left(1 + \dfrac{R_4}{R_5}\right)}{\left(\dfrac{1}{R_1} + \dfrac{1}{R_2} + \dfrac{1}{R_3}\right)} = K_1V_1 + K_2V_2$$

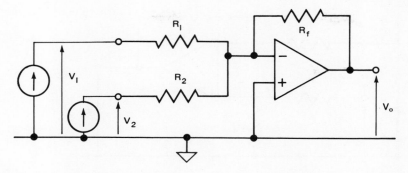

Fig 12.4 Operational amplifier network providing summation and sign inversion

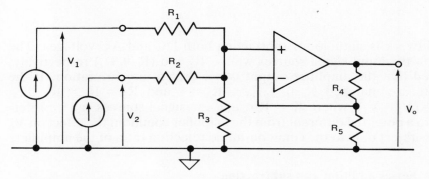

Fig 12.5 Operational amplifier network providing summation without sign inversion

This network is suitable for adding both DC and AC voltages. The loads on the input signal sources will be approximately

$$\left(R_1 + \frac{R_2 R_3}{R_2 + R_3}\right) \quad \text{and} \quad \left(R_2 + \frac{R_1 R_3}{R_1 + R_3}\right) \text{ respectively}$$

irrespective of whether the amplifier is saturated or not. Both input signal sources need to be single-ended and referenced to common. R_3 may be omitted physically and replaced by infinity in the above expressions, which then both reduce to $(R_1 + R_2)$.

12.2.3 Differential amplifier configuration

Referring to Fig. 12.6, it is shown readily that

$$V_o = -\frac{R_3}{R_1} V_1 + \frac{R_4}{(R_2 + R_4)} \cdot \frac{(R_3 + R_1)}{R_1} V_2 = -K_1 V_1 + K_2 V_2$$

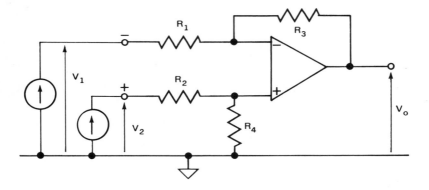

Fig 12.6 Operational amplifier network providing differential inputs

This network is suitable for subtracting both DC and AC voltages. The loads on the input signal sources will be R_1 and $(R_2 + R_4)$ respectively, provided that the amplifier output is not driven into saturation. If the resistors are matched so that $R_1 = R_2 = r$ and $R_3 = R_4 = R$, then $V_o = K(V_2 - V_1)$, where $K = R/r$, and the signal sources can be referenced to a potential different from the amplifier common: the effect on V_o will be subject only to the common-mode rejection ratio of the amplifier.

12.2.4   Series addition and subtraction

Referring to Fig. 12.7, it is readily shown that

$$V_o = \left(1 + \frac{R_1}{R_2}\right)(V_1 + V_2) = K(V_1 + V_2), \qquad K > 1.$$

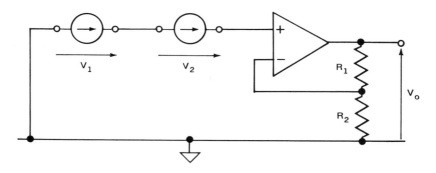

Fig 12.7 Operational amplifier network providing for series addition and subtraction of voltages

This network is suitable for adding both DC and AC voltages. The loads on the input signal sources will be almost infinitesimal, being due to the (very high) input impedance of the amplifier and to the (very low) input offset current of the amplifier. The first signal source must be single-ended and therefore connected to common, whilst the second (and any addition-al) signal source must be floating. The network can be modified to subtract signals, by reversing one of the signal sources, where appropriate.

Where the range of V_1 and/or V_2 is too great for the amplifier, these signals may be attenuated prior to being mixed, as shown in Fig. 12.8. In this example

$$V_o = \left(\frac{R_2}{R_1 + R_2}\right)V_1 + \left(\frac{R_4}{R_3 + R_4}\right)V_2$$

$$= K_1 V_1 + K_2 V_2 \qquad K_1 < 1, K_2 < 1$$

The loads on the signal sources are now $(R_1 + R_2)$ and $(R_3 + R_4)$, respectively.

12.2.5   Analog comparators

The purpose of an *analog comparator* is to generate a digital signal to indicate the relative state of two analog signals. In theory, at least, any of the networks featured in Figs 12.4 to 12.7 inclusive may be configured to this end, by choice of appropriate values for specified resistors, as follows:

Network Fig. no.	Necessary resistor values	Input state for $V_o = V_{sat_-}$	Input state for $V_o = V_{sat_+}$
12.4	$R_f = \infty$	$(V_1 + V_2) > 0$	$(V_1 + V_2) < 0$
12.5	$R_5 = 0$	$\left(\dfrac{V_1}{R_1} + \dfrac{V_2}{R_2}\right) < 0$	$\left(\dfrac{V_1}{R_1} + \dfrac{V_2}{R_2}\right) > 0$
12.6	$R_3 = \infty$	$\left(\dfrac{R_4 V_2}{(R_2 + R_4)} - V_1\right) < 0$	$\left(\dfrac{R_4 V_2}{(R_2 + R_4)} - V_1\right) > 0$
12.7	$R_2 = 0$	$(V_1 + V_2) < 0$	$(V_1 + V_2) > 0$

V_{sat_+} and V_{sat_-} are the output saturation levels of the operational amplifier, the values of which will depend upon the values of the amplifier supply rail voltages. For high-speed switching, it is not good practice to drive the output voltage into saturation, but to limit its excursion by means of appropriate components: Fig. 12.9 shows two typical arrangements.

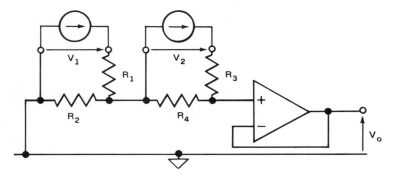

Fig 12.8   Operational amplifier network providing for series addition and subtraction of voltages in attenuated proportions

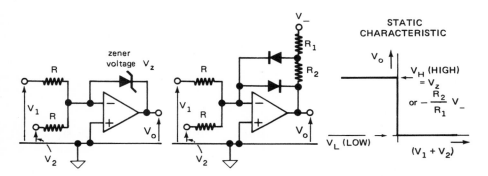

Fig 12.9   Two alternative simple operational amplifier networks providing for comparison of voltages (analog comparators)

Far superior in performance to these networks are dedicated analog comparator ICs, such as the LM311, which offer the following advantages:

- relatively high differential input voltage range
- unipolar supply rail voltage requirement
- low response time (200 ns is typical)
- very high input impedance
- low input offset current
- open collector output.

The advantage with the open collector output arises from the fact that the HIGH level V_H of output voltage can be chosen to suit the application, with an appropriate 'pull-up' resistor being connected between the output terminal and a suitable V_H supply rail. Note that most analog comparator ICs respond to an input voltage differential, as opposed to an input voltage sum.

12.2.6   Use of bridge networks for subtracting signals

In the arrangement shown in Fig. 12.10, which is, in effect, an unbalanced Wheatstone bridge, it can be shown that $V_o = K(\theta_i - \theta_o)$, where K is the V/rad sensitivity of the feedback potentiometer. The network performs equally well for DC and AC signals. Only one point in the network supplying the amplifier can be connected to common: either one amplifier input can be commoned or one side of the supply can be commoned, but they must not both be commoned, because this will short-circuit one limb of the bridge. The formula for V_o is only precise provided that the potentiometers are not loaded significantly.

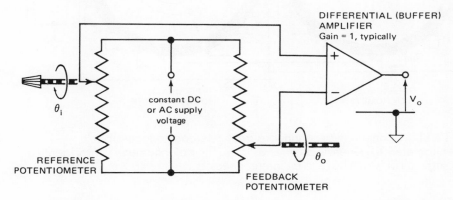

Fig 12.10   Connection of reference and feedback potentiometers to generate an error voltage

12.2.7   Use of tapped continuous track potentiometers for subtracting signals

A development of the previous method employs a pair of special toroidal servo potentiometers each having a continuous track, multiple equi-spaced tappings, and dual brushes set diammetrically opposite to one another. Each of the set of tappings on one potentiometer is joined to the corresponding tapping on the other, as indicated in Fig. 12.11: most of the links have been omitted for clarity. Using slip rings (not shown), a fixed DC or AC voltage, usually from a floating source, is applied to the two brushes of the reference potentiometer. This causes two identical linear voltage distributions to be established along the two 180° sectors of the track. The alignment of these two distributions will be determined by the value of θ_i selected. The voltage distributions are detected by the tappings and are transferred to the tappings on the feedback potentiometer. The resulting voltage distributions imposed upon the track of the feedback potentiometer will be almost a replica of those on the reference potentio-meter, the accuracy of reproduction depending upon the number of

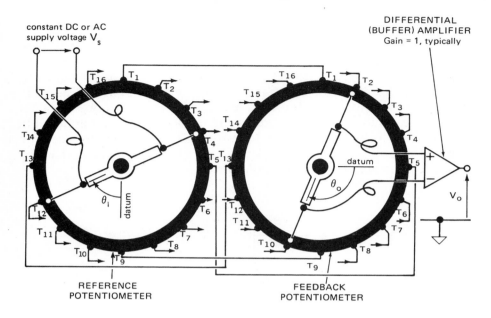

Fig 12.11   Connection of continuous track reference and feedback potentiometers to generate an error voltage. The potentiometers carry identical tappings, all of which have to be interconnected

tappings. In any case, when $\theta_o = \theta_i$ the two brushes of the feedback potentiometer will be in contact with two points on the track at identical potentials, so that the output potential difference, which is transferred to the amplifier via slip rings (again not shown), will be zero. Note that θ_i and θ_o are calibrated relative to datums separated by 90°. When $\theta_o \neq \theta_i$, it is shown readily that

$$V_o \cong V_s \cdot \frac{2}{\pi}(\theta_i - \theta_o) \qquad \text{when} -\frac{\pi}{2} < (\theta_i - \theta_o) < \frac{\pi}{2}$$

and

$$\cong -V_s \cdot \frac{2}{\pi}(\theta_i - \theta_o) \qquad \text{when} - \pi < (\theta_i - \theta_o) < -\frac{\pi}{2}$$

$$\text{and when} \quad \frac{\pi}{2} < (\theta_i - \theta_o) < \pi$$

V_s is the supply voltage and $(\theta_i - \theta_o)$ is in radians.

12.2.8 Use of electromagnetic fields for adding and subtracting signals

An alternative method for combining data represented by electrical signals is to cause these signals to establish magnetic fields, to combine these fields, and to sense the value of the nett flux, the output signal normally being a voltage. A typical technique is to employ multiple field windings on a DC generator, which is being used as a power amplifier, as shown in Fig. 12.12. The separate field windings provide full isolation

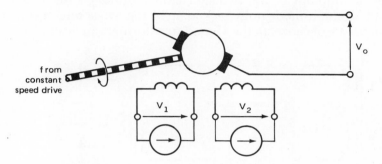

Fig 12.12 Use of multiple windings on magnetic circuits to achieve addition and subtraction of analog signals

between the input signal sources. Reversing a signal source converts additive action to subtractive action, and vice-versa. Assuming a linear magnetisation characteristic, the configuration shown in Fig. 12.12 will yield an expression of the form $V_o = K_1 V_1 + K_2 V_2$, where the relative values of K_1 and K_2 will depend upon the relative numbers of turns and the relative resistances of the two windings.

The technique which has been described here is suitable only when the signal voltages are DC. When the sources are AC, transformer action between windings having a common magnetic circuit would result in each AC voltage source inducing high circulating currents in the other input circuits, which would not be tolerable.

The use of synchro differentials and control transformers in chains of synchros, as described in Section 3.2.5, presents an electromagnetic technique for the addition and subtraction of angular data. Resolvers, discussed in Section 3.2.6, provide an alternative set of components which can be applied in similar fashion. In these cases, all of the electrical signals are AC voltages.

12.2.9 Additional techniques with AC signals

One of the advantages with using AC signals arises from the fact that signal transformers can be introduced into the signal paths, prior to signal combination. Signal transformers can provide some or all of the following facilities:

- electrical isolation of signal sources from loads
- signal level and impedance level conversion
- improvement in signal-to-noise ratio.

For an example of the application of signal transformers, see Fig. 12.13. The centre-tapped windings of the two signal transformers form a balanced bridge and common-mode noise picked up by the two conductors in the transmission path will cancel in the second transformer, and therefore

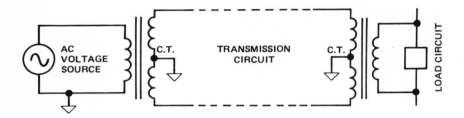

Fig 12.13   Application of signal transformers in transmission paths of AC signals

will not be transmitted to the load. In addition, the two conductors may be twisted together and the twisted pair may be screened, with the screen connected to common (ground). Electrically, the transmission circuit is completely symmetrical relative to common. The signal source and the load are completely isolated from each other electrically, the signal source need only be single ended, and the transformer turns ratios can be selected for signal level and impedance level requirements.

12.3   Electronic networks for combining digital data

The application of arithmetic operations to digital data is most easily undertaken when the data are represented in some form of weighted code, as defined in Section 3.2.8. Unless otherwise stated, the operations to be described here assume that a natural binary code is used during the processing, and this is by far the most common format: however, the techniques may be adapted to accommodate other weighted formats. The networks to be described involve the application of hard wired logic to arithmetic operations, although a suitable digital processor may be programmed to perform the same tasks.

12.3.1   Natural binary representation

With a word length of eight bits, for example, all eight bits can be allocated to the representation of a numerical value, if the value always has the same polarity. In those cases where the polarity can assume either sign, one of the bits must be used to indicate the sense of the polarity, and thus becomes the 'sign bit'. The most common convention used is one which allocates, to the most significant bit, binary 0 to represent a positive value and binary 1 to indicate a negative value.

Thus, a positive integer might be represented by the word 0100 1101 which, according to the basis for the discussion so far, is to be interpreted as

$$+ [(1 \times 2^6) + (0 \times 2^5) + (0 \times 2^4) + (1 \times 2^3) + (1 \times 2^2) + (0 \times 2^1)$$
$$+ (1 \times 2^0)] = + 77_{10}$$

In order to facilitate arithmetic operations, the magnitude of a negative value of data would usually be represented in 'complement' form, the most common being 'two's complement format'. The two's complement of a number is formed by inverting every digit and then adding a 1 to the result: thus, to encode -77_{10} one would need to invert 0100 1101, yielding 1011 0010, and then add 1, so that the final code is 1011 0011.

In those cases where a fractional number is required to be represented, the binary digits now acquire weightings which are negative powers of the base 2. Thus, a positive fraction might also be represented by the word 0100 1101 which, using a comparable basis for interpretation, would denote a value

$$+ [(1 \times 2^{-1}) + (0 \times 2^{-2}) + (0 \times 2^{-3}) + (1 \times 2^{-4}) + (1 \times 2^{-5})$$
$$+ (0 \times 2^{-6}) + (1 \times 2^{-7})] = + 0.601\,562\,5_{10}$$

The two's complement technique can also be applied to the representation of negative fractions, so that $-0.601\,562\,5_{10}$ would be coded by inverting every digit of the word 0100 1101, yielding 1011 0010, and then adding 1 to the least significant bit, to produce 1011 0011, as for the integer case.

Arithmetic operations are much more easily undertaken using integer numbers alone or fractional numbers alone and, wherever possible, digital control hardware should attempt to scale the data to this effect. Where the range of values to be processed cannot be accommodated in this manner (not even with the use of multiple length words), mixed numbers must then be used, in which case the representation normally adopted is 'floating point format'. In this format, a mixed number typically is represented in the form $M \times 2^E$, where M is a signed fractional 'mantissa' (normally adjusted to lie between 0.5_{10} and 1 in magnitude) and E is a signed integer 'exponent'. Multiplication and division of two floating point numbers $M_1 \times 2^{E_1}$ and $M_2 \times 2^{E_2}$ (say) are relatively easy operations, because

$$(M_1 \times 2^{E_1}) \times (M_2 \times 2^{E_2}) = (M_1 \times M_2) \times 2^{(E_1 + E_2)}$$

and

$$(M_1 \times 2^{E_1})/(M_2 \times 2^{E_2}) = (M_1/M_2) \times 2^{(E_1 - E_2)}$$

Thus, these two types of operation involve the multiplication and division of fractional numbers and the addition and subtraction of integer numbers. Addition and subtraction of two floating point numbers require that the two numbers have the same value of exponent, so that provision must be made for adjusting exponents and mantissae accordingly, before processing can proceed further: thus, for example

$$(+0.8 \times 2^{+10}) + (-0.6 \times 2^{+8}) = (+0.8 \times 2^{+10}) + (-0.15 \times 2^{+10})$$
$$= (+0.8 - 0.15) \times 2^{+10} = +0.65 \times 2^{+10}$$

After any processing operation, including multiplication and division, it will be usual to adjust, where necessary, the mantissa and exponent of the result, so that the former finally lies within the range 0.5_{10} to 1 in magnitude.

The addition of two signed integer numbers can follow the rules of binary addition, with the sign bit handled in the same manner as any other bit. For example

Decimal	Natural binary
+16	010000
+(−3)	+111101
+13	(1)001101

↑ ↑ sign bit

discard carry bit

Subtraction of a signed integer number is usually implemented by firstly complementing the number to be subtracted (the 'subtrahend') and then adding the result to the 'minuend'.

In the addition and subtraction processes described here, any leftmost carry bit is always discarded, provided that two's complement arithmetic is used. Thus, for example:

Decimal	Natural binary
+3	000011
−(+16)	+110000
−13	(0) 110011

↑ sign bit

zero carry bit

Decimal	Natural binary
−29	100011
−(+2)	+111110

| −31 | (1)100001 |

↑ ↑ sign bit

discard carry bit

The need for multiplication and division operations in digital control processes may arise where it is required to apply a gain constant to a variable or, in rarer cases, where two variables need to be multiplied or divided.

The multiplication of two signed integer numbers can follow the rules for binary multiplication, provided that the sign bits are processed separately. For example

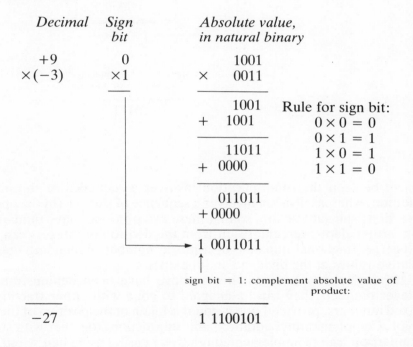

Decimal	Sign bit	Absolute value, in natural binary	
+9	0	1001	
×(−3)	×1	× 0011	
		1001	Rule for sign bit:
		+ 1001	$0 \times 0 = 0$
			$0 \times 1 = 1$
		11011	$1 \times 0 = 1$
		+ 0000	$1 \times 1 = 0$
		011011	
		+ 0000	
		1 0011011	

↑

sign bit = 1: complement absolute value of product:

| −27 | 1 1100101 |

It will be seen that the word length of the product is double that of the multiplicand and the multiplier, and provision would need to be made for this situation.

The division of two signed integer numbers can follow the rules for binary division, provided that the sign bits are handled separately. For example

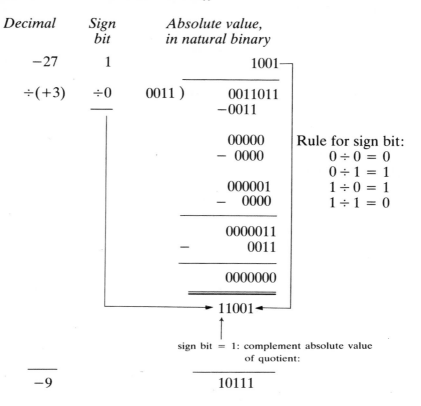

It can be seen that multiplication involves a sequence of shifting and addition, whereas division involves a sequence of shifting (in the opposite direction) and subtraction, so that these two processes have fundamental similarities. However, the result from the division of integers can be an integer, a fractional number, or a mixed number, depending upon the relative values of the dividend and the divisor.

Although the operations shown above have been demonstrated for integer numbers they can be adapted to cope with either fractional or mixed numbers, provided that account is taken of the position of the radix point. Complementing, addition, and subtraction, together with straight comparison, can be implemented relatively easily, using hardwired logic; although such implementation is also possible for multiplication and division, the complexity involved makes these last two processes more easily handled with a programmable digital processor, with the arithmetic operations specified in terms of program statements.

12.3.2   Hardwired logic for forming the two's complement

When an n-bit natural binary word has to be translated into two's complement form, a process representing the sign inversion of data, a suitable algorithm for the jth bit is $B_j = A_j \oplus [A_{j-1} + A_{j-2} + \ldots + A_1 + A_0]$, $j > 1$, and $B_0 = A_0$, where $A_{n-1} \ldots A_0$ are the bits of the word before it is converted and $B_{n-1} \ldots B_0$ are the corresponding bits of the word after conversion. This algorithm is valid whether A_{n-1} and B_{n-1} represent sign bits or not. The symbol $\oplus$ signifies the Exclusive – OR (modulo-2 addition) process. Figure 12.14 shows one possible logic network for implementing this algorithm for an 8-bit word.

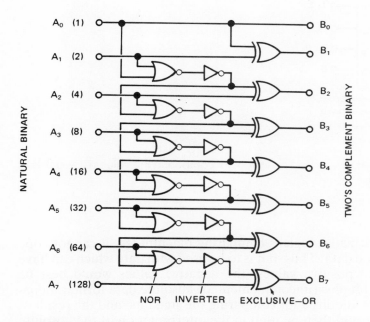

Fig 12.14   Typical hardwired logic network for generating the two's complement of an 8-bit natural binary word

12.3.3   Hardwired logic for comparing two binary words

The logic to be described here represents the digital counterpart of the analog comparator of Section 12.2.5, and so could be used in the formation of a digital on-off controller, for example. Medium-scale ICs are available for generating a comparison between two digital words of limited length. An example in the TTL range is the 7485, which can be used to compare two 4-bit words: these ICs can be cascaded, in order to compare two natural binary or B C D words having lengths which are integer multiples of four bits, as shown in Fig. 12.15.

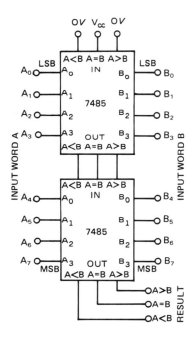

Fig 12.15 Typical hardwired logic network for comparing the values of two 8-bit natural binary words

Because this particular logic is comparing magnitudes, it cannot cope with complemented data. Thus, if it is to be used with data which can have both negative and positive values, the negative values would best be converted to sign bit plus (uncomplemented) absolute data format. Separate logic would be required for comparing the sign bits, and the result of this comparison would then be used in the interpretation of the absolute value comparison.

Any unused IC data inputs would be connected to logic LOW.

12.3.4 Hardwired logic for adding two natural binary words

The logic to be described here represents the digital counterpart of the non-inverting analog summer of Section 12.2.2, scaled for unity gain. Medium-scale ICs are available for generating the sum of two binary words of limited length. An example in the TTL range is the 7483, which can be used to add together two words having lengths which are integer multiples of four bits, as shown in Fig. 12.16. This particular logic is designed to handle either unsigned or signed-and-complemented data, so that no special treatment is required for the latter case. Any unused IC data inputs should be connected to logic LOW.

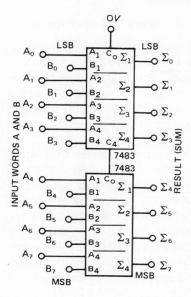

Fig 12.16 Typical hardwired logic network for adding two 8-bit natural binary words

12.3.5 Hardwired logic for multiplying two 4-bit natural binary words

Figure 12.17 shows two particular ICs which can be connected to generate the 8-bit product of two 4-bit natural binary words. The configuration represents a (limited) hardwired logic arrangement for multiplying unsigned 4-bit words. For handling data of any greater complexity, a digital processor would probably be used.

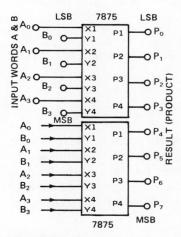

Fig 12.17 Typical hardwired logic network for multiplying two 8-bit natural binary words

12.3.6   Logic for performing more complex mathematical operations

Hardwired logic may be designed to handle the addition, subtraction, multiplication, division and comparison of two's complement floating point data, for example, using techniques based upon those described in Sections 12.3.2 to 12.3.5. However, the recent rapid reduction in the cost of VLSI general-purpose processors and co-processors and special-purpose digital signal processors would make this approach uneconomical, apart from the most exceptional circumstances.

The ready availability of these VLSI devices now means that most digital signal processing will, in the future, be undertaken either in software, in hardware, or in a combination of software and hardware. Generally speaking, software processing involves a rather lower cost of hardware but requires more programming complexity and/or software cost and results in relatively long execution times. Conversely, hardware processing is rather more expensive, requires reduced programming complexity and software outlay (because many of the mathematical routines are supplied by the VLSI manufacturer), and achieves very fast rates of execution.

This approach to digital signal processing is beyond the scope of this book: it will, however, be covered in detail in a companion volume.

12.4   Mechanical methods for combining signals

Mechanical devices can be used to generate the sum and difference between linear displacements or forces, for the rectilinear case, and between angular displacements or torques, for the rotary case. The advantages of these devices are their simplicity, ruggedness and independence from power supplies. Possible disadvantages include:

- friction, mechanical hysteresis, inertia, and deflection (bending) effects
- bulkiness
- inflexibility of the configurations, in contrast to electronic networks.

Some typical mechanical devices will be described.

12.4.1   The lever and the walking beam

In the application shown in Fig. 12.18, the *walking beam* is a lever which effects a mechanical negative feedback path around an hydraulic servo-valve-power cylinder combination. The quiescent position of the lever is indicated by the reference axis. When e = 0, the valve spool is central, so that no oil flows into the cylinder and the piston will be stationary. Applying similar triangles

$$\frac{y}{b} = \frac{x}{a}$$

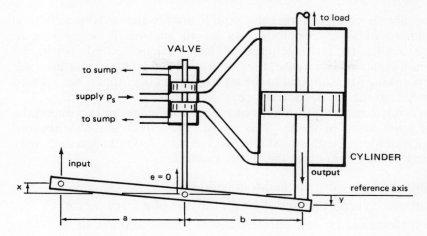

Fig 12.18 Use of a lever and hydraulic servovalve–power cylinder combination to produce a walking beam

If the input x is changed, the beam pivots about the output joint at y, because the large holding forces initially prevent the piston from moving. This results in e changing, and the displaced valve spool causes the piston to be driven until the subsequent change in y has reduced e back to zero. This condition corresponds to a new equilibrium state.

For small variations about the quiescent position

$$\Delta e = \frac{\partial e}{\partial x} \cdot \Delta x \;+\; \frac{\partial e}{\partial y} \cdot \Delta y$$

$\partial e/\partial x$ can be derived from the relationship between e and x with the output end unmoved: using triangles

$$\frac{\partial e}{\partial x} = \lim_{\substack{\Delta e \to 0 \\ \Delta x \to 0}} \left[\frac{\Delta e}{\Delta x} \right] = \frac{b}{(a + b)}$$

By similar reasoning, for the other end

$$\frac{\partial e}{\partial y} = - \frac{a}{(a + b)}$$

Thus

$$\Delta e = \frac{b}{(a + b)} \cdot \Delta x \;-\; \frac{a}{(a + b)} \cdot \Delta y = K_1 \Delta x - K_2 \Delta y$$

With the above configuration, the equilibrium state corresponds to a 'displacement balance' condition. As an alternative, it is possible to configure the hardware to generate a 'force balance' condition in the equilibrium state: in such a case, the output will deflect by an amount such that the resulting force applied to the load balances the input force applied to the beam, in the proportion of a:b.

By applying forces to different points along the beam, and reversing the sense of application of forces, it is possible to effect the addition and subtraction of forces, with adjustable proportions. A controlled force may be applied to the beam input using, for example:

- a controlled pressure applied to a bellows
- a controlled current passed through the coil of a force motor
- a spring extended by a controlled displacement of one end.

These techniques form the basis of differential–pressure–to–pressure and differential–pressure–to–current transmitters (Section 5.2) and control valve positioners (Section 10.1.4). The advantages arising from using the force balance technique are that effects such as those due to mechanical hysteresis, deadband, and static friction are minimised.

12.4.2   The differential gear

Angular displacement may be added or subtracted (depending upon the sense of the displacement) using the type of differential gear shown in Fig. 12.19. If there is relative motion between the two input shafts, so that $\theta_1 \neq \theta_2$, then the axis of gear G_1 will be rotated bodily about IOI in a plane through POP and perpendicular to the page. This will result in rotation of gear G_2 which will be transmitted, through gear G_3, to the output shaft. $\theta_o = K(\theta_1 + \theta_2)$, where the value of K depends upon the gear ratios.

12.5   Gear trains

In many servomechanisms, it is commonplace to use one or more gear trains in the vicinity of the servomotor (or actuator), the mechanical load, and the feedback transducer. There are many possible reasons for this, amongst which feature the following:

- to enable a high-speed–low-torque motor (or actuator) to be coupled to a low-speed–high-torque load, in order to minimise motor (or actuator) size
- to convert rectilinear motion to rotational motion, and vice-versa, using rack-pinion or screw-nut configurations
- to reverse the direction of motion of a drive
- to provide a change in the direction of a shaft drive (for example, to turn a drive through 90°)

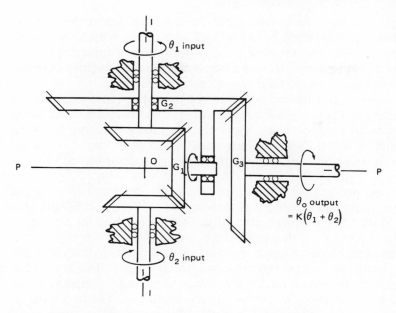

Fig 12.19 Use of a differential gear to provide addition and subtraction of shaft angular displacements

- to enable a transducer to be coupled to an otherwise inaccessible shaft, the selection of gear ratio providing means for mechanically altering the system loop gain

- to enable shaft angular displacements to be added or subtracted (see Section 12.4).

In the last two cases, the gears are usually transmitting data at very low power levels and, as such, would be termed 'instrument gears'. Instrument gear trains are available commercially in pre-packaged form, covering a wide range of alternative gear ratios; often they are packaged in synchro-style cases and then are known as 'gearheads'.

It is possible to make backlash-free instrument gears, by 'splitting' pinions into two parallel plates. These are sprung apart, in an angular sense, so that one is in contact with the 'leading' edges of the teeth of the mating gear, whilst the other is in contact with the 'trailing' edges, so that no out-of-mesh state can ever exist. Against the advantages of backlash-free gearing must be traded the disadvantages of added complexity and a significant increase in frictional forces.

12.6 Brakes, clamps and clutches

In a number of servosystem applications, there may be a requirement for the drive to be capable of holding the load rigidly stationary, against the action of reactive load forces. One example would be when the drive is hoisting a load and must then hold the load at a given height. A second example would be in a machine-tool drive, where the workpiece and table must be held rigidly stationary during a machining operation (for example, during a milling cut or a gear hobbing operation).

Electric motors, in particular, require to be in motion, in order that the ventilating action can provide adequate cooling, if the motor is developing significant levels of torque. Thus, most electric motors are not rated to sustain significant torque levels when crawling or stationary: to be so would necessitate a significant increase in frame size and/or forced ventilation.

Hydraulic and pneumatic drives can usually be designed to have an inherent capacity for developing high holding forces/torques. Where this requirement is specified for an electric drive, it usually becomes necessary to supplement the motor with a brake or clamp.

In certain motors, electromagnetically actuated brakes are built into the motor frame, as an integral part of the armature assembly. If such a motor is not available, it becomes necessary either to include a brake in the shaft-gearing-load assembly or physically to clamp the load to a fixture.

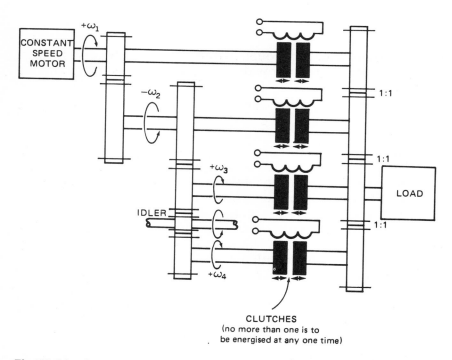

Fig 12.20   Symbolic representation of the application of a gear train and electromagnetic clutches to produce a set of alternative load speeds and directions of rotation

Brakes and clamps, under automatic control, would be actuated by either electromagnetic, hydraulic, or pneumatic actuators.

Clutches can be used, in an on-off manner, in 'bang-bang' speed control systems. A typical arrangement is shown symbolically in Fig. 12.20. In this arrangement, with the motor assumed to be run at constant speed $+\omega_1$, the load could be driven at alternative speeds of $+\omega_1$, $+\omega_2$, $+\omega_3$, $-\omega_4$, depending upon which clutch is energised. With no clutch energised, the load would be stationary (and might require to be clamped). Such on-off clutches may be electromagnetically, hydraulically, or pneumatically actuated.

It is also possible to create a (unidirectional) variable speed drive using a constant speed motor, if a coupling capable of controllable slip is introduced. (Note that normal clutches would burn out if allowed to slip for any length of time). One possible coupling for this application is the type of fluid flywheel coupling used in automotive transmissions. An alternative is an electrical equivalent, one of which is the 'magnetic particle clutch': the two clutch faces are separated by a 'plasma' of magnetic particles, and the degree of slip between the two faces is determined by the degree of magnetisation to which the particles have been subjected. This is controlled by manipulating the DC current passed through a stationary solenoid. Another electrical alternative is the 'eddy current clutch', which is based upon a metal disc being rotated within the field of a rotating electromagnet, with manipulation of the current being passed through the winding: the currents generated within the disc react with the magnetic field, to produce a (controllable) coupling torque and controllable slip.

13

Development of Complete Systems and the Construction of Schematic Diagrams

13.1 Development of complete systems

Experience shows that the control engineer is rarely consulted in the initial development stages, when a new plant is under construction. Thus, he is usually presented with a completed plant design, as a *fait accompli*, and is requested to control it. Obviously a prerequisite for the control engineer is that he should have a reasonable knowledge of the operating principles and the technical details of the plant, before he can attempt to control it with an adequate level of performance.

His first task is to identify those properties (the controlled variables) of the plant which are ultimately to be controlled. Having done so, he can formulate a specification for the degree of precision to which each variable is to be controlled, both in a dynamic sense and in a steady state sense, taking into account the parasitic disturbances to which the plant is likely to be subjected and the acceptance limits on the quality of whatever product is being produced by the plant.

His next task is to identify those properties (the manipulated variables) of the plant which are going to be manipulated, by the final control elements, in order that the variations in the controlled variables shall remain within specification. Usually, the choice of manipulated variables is limited and, moreover, obvious.

Having identified the controlled variables, these must be instrumented with appropriate feedback transducers. Ideally, each controlled variable would be measured directly, by means of its allotted transducer. In rare instances, no suitable transducer may exist or the variable may be inaccessible (due, for example, to a harsh environment), in which case it will be necessary to transduce related variables and to infer, from the signals generated, the value of the controlled variable of interest, using appropriate computing hardware.

The sorts of factor which could be relevant to the selection of a feedback transducer have been listed in Section 1.5, but will be repeated here for completeness:

- cost

- availability
- ruggedness, in respect to the plant environment
- range
- accuracy
- linearity
- repeatability
- speed of response
- reliability
- maintainability
- life
- power supply requirements
- physical compatibility with the plant
- signal compatibility with the controller
- signal-to-noise ratio.

Having identified the manipulated variables, it becomes necessary to select final control elements, in order to provide the necessary manipulation. In most cases, these will be either control valves, heaters, pumps, actuators, or motors and the factors relevant to motor selection (for example) were listed in Section 1.5 and are repeated here:

- cost
- availability
- ruggedness, in respect to the plant environment
- load details: inertia, friction constants, torque loadings
- maximum and minimum velocity
- maximum acceleration
- duty cycle
- reliability
- maintainability
- life
- mounting and coupling requirements
- power supply requirements
- input signal characteristics.

The feedback transducer and final control element in each loop are physically connected to the plant, and therefore have to be selected to cope with the plant environment. The remaining components in each control loop are not physically coupled to the plant, being linked to the feedback transducer and final control element usually by means of electrical, pneumatic or hydraulic connections, and may even be sited quite remotely from the plant: often, these remaining elements will be situated in a control room, which will usually have a much more benign environment than that of the plant.

The remaining elements which would normally constitute each analog loop perform the following functions:

- signal and power amplification of the error signal, in order to drive the final control element
- means for generating reference data, to be manually set in by the operator or electronically set by computing hardware
- hardware for generating the error signal from the reference and feedback signals
- hardware (for example, active R-C filters) for modifying the dynamic and steady state behaviour of the loop, in order to satisfy the performance specification: this is referred to as 'compensation'.

In the case of process loops, all of these functions can be provided by the general-purpose process controller which is normally used in such cases. In other instances, it is often necessary for the control engineer to custom-design a controller to perform these tasks. In the case of digital controllers, most or all of these functions (except power amplification) would be synthesised digitally.

In many cases, ancillary equipment may be required to perform one or more of the following tasks:

- incorporation of additional feedback or feedforward signals from additional transducers or controllers
- signal conversion and/or conditioning
- alarm indication when system performance exceeds preset limits
- limitation of the reference variable values, in terms of (say) magnitude and/or rate of change
- limitation of the controller output or final control element output action, to prevent (for example) unnecessary stressing of the plant
- noise filtering
- automatic detection and indication of equipment failure
- sequence control, for scheduling the sequence of operating levels and modes of the control loop.

As a final stage of commissioning the hardware of each control loop, it is usually necessary to adjust (that is, 'fine tune') the compensation elements, in order to optimise the dynamic and steady state performance. When general-purpose process controllers are being used, there is a range of alternative procedures available for tuning the numerical values of the terms in the control law, and these are usually based upon simple tests and routine adjustment procedures. When a controller is being custom designed, a much more elaborate procedure is necessary: this often involves identifying (characterising) the dynamic and steady state properties of the plant process, between the input to the final control element and the output from the feedback transducer, and then designing the optimum control law, using one or more a number of alternative design techniques; finally, the completed system must be tested in order to demonstrate that it complies with the performance specification. Many of these topics will be covered in detail in the companion to this volume.

13.2 Construction of schematic diagrams

The diagrammatic representation of a complete control system is often made in terms of what is generally known as a 'schematic diagram'. However, there is no specific definition of exactly what constitutes such a diagram. Certainly, a schematic diagram is distinguishable from a block diagram: in the latter case arrowed interconnections indicate data flow and each block represents a mathematical relationship between output data and input data.

The following comments would apply generally to schematic diagrams:

- all physical interconnections (including signal common connections, in electrical cases) which transmit signals representing system data transfer would be shown

- power supplies and power supply connections would not usually be shown

- terminal identifiers would often be added for the signal interconnectors, as an aid to fault finding

- high power level circuits (electrical, hydraulic and pneumatic) usually would be shown in detail

- mechanical elements (gears, clutches, couplings, etc.) would be shown and represented by appropriate pictorial symbols

- the plant process would be included and represented by appropriate pictorial symbols

- complex low power level circuits (for example, electronic amplifiers, electronic or pneumatic logic networks, etc.) would often be represented by a simple pictorial symbol, with the details supplied elsewhere on other diagrams

- transducers would be represented by appropriate pictorial symbols

- analog hardware involved in signal combination (addition and subtraction) would typically be shown in detail.

The completed schematic diagram thus yields a wealth of information about the physical interrelationships between the different system elements which constitute each control loop. Such a diagram can be regarded as a stage in the process of analysing system behaviour, because usually the block diagram for the loop can be developed readily from the schematic diagram, certainly in terms of the configuration of the block diagram: the schematic diagram will yield little information about the transfer functions and static characteristics required for the block diagram, except that their general form may be implied.

13.3   Examples of the development of a control system and its schematic diagram

The evolution of this system will follow a natural sequence.

1 Formulation of the requirements, in qualitative terms.
2 Specification of the ultimate performance in quantitative terms.
3 Selection of system hardware.
4 Development of a schematic diagram.
5 Design, construction and commissioning of the system.
6 Identification of the dynamic and steady state characteristics of the system hardware.
7 Fine tuning of the control laws.
8 Demonstration that performance meets specification.

The example to be presented is hypothetical, so that only stages 1, 3 and 4 in the above procedure will be described in detail.

13.3.1   Requirements for the example system

The requirements are for a precision linear position control system for a machine tool. The degree of accuracy and resolution is to be in the order of 0.001 mm over a total travel of (say) 1 metre, which amounts to 1 part in 10^6. The system is to have a very fast dynamic response and is to be capable of slewing at rates up to (say) 50 mm/second. There must be no relative motion between the workpiece and cutting tool whilst cutting is taking place, so that the system is required to generate high holding forces in order that the cutting reaction does not generate significant displacement errors.

The above requirements are, to a large extent, qualitative and have been expressed in fairly general terms. In practice, a detailed performance specification in quantitative terms would now have to be formulated.

13.3.2   Selection of the feedback transducer

To obtain the degree of accuracy and resolution required, the selection of a feedback transducer would have to be limited to the following alternatives:

- optical diffraction grating or equivalent
- Inductosyn, in a coarse-fine measuring system
- multi-turn shaft encoder containing two encoding discs geared together internally

- multiple synchro chains geared together, to produce a coarse-fine measuring system.

Let us suppose that, having applied the factors listed for transducer selection in Section 13.1 to commercially available devices, it has been decided to use an optical diffraction grating. This device with its integral electronics will generate pulses, to indicate increments in displacement, together with signal(s) representing the direction of motion. These signals must be input to a reversible up/down counter, which will generate a count representing the displacement of the machine tool slide from an arbitrary datum. Means must be included to enable the counter to be reset manually to zero, at the requirement of the operator.

13.3.3   Reference and error data generation

The count size will be available as a set of logic signals, in a suitable binary code, representing controlled variable data. Because it is not practicable to generate analog signals to an accuracy approaching 1 part in 10^6, it is necessary to set up the desired position (the reference data) as a compatible digital word. The error data can than be computed digitally, using the reference and feedback words, by means of hardware to effect binary subtraction (or addition, by suitable manipulation). The error data can then be converted to a simple analog electrical signal, to drive the power amplifiers, by means of a suitable digital-analog converter.

13.3.4   Selection of the final control element

To obtain the velocity, acceleration, and holding force levels required, it is almost certain that a hydraulic drive can satisfy the requirements. The final control element could be a rotary hydraulic piston motor, with the rotary motion being converted to rectilinear by means of a leadscrew and nut: the backlash that could be introduced between the screw and nut would degrade system performance, but this could be minimised by using (say) a recirculating ball-bearing type of arrangement, which is expensive but possesses considerably diminished backlash levels.

An alternative would be to generate rectilinear motion directly, by virtue of using a hydraulic cylinder as the final control element: backlash would not arise but the length of travel might be excessive for commercially available cylinders; in addition, a cylinder is likely to give a slower response than a piston motor would.

Let us suppose that, having applied the factors listed for final control element selection in Section 13.1 to commercially available motors and cylinders, it has been decided to use a hydraulic piston motor, in conjunction with a recirculating ball leadscrew and nut. This motor will then need to be controlled, by means of manipulation of the flow of high-pressure hydraulic fluid into, and out of, the motor.

13.3.5   Selection of the amplifiers

The performance specification for this type of system will necessitate a very high level of loop gain, and most of this loop gain must be provided by appropriate levels of amplification in the forward path of the loop. In addition, the requirements of the motor would necessitate a relatively high level of power amplification. Therefore, a series of cascaded amplifier stages would almost certainly be required, the combination of which would provide the levels of signal and power amplification necessary.

A typical arrangement for the type of system under consideration would be a hydraulic spool type of servovalve to control the motor flow, with the spool being displaced by the hydraulic pressure controlled by a flapper-nozzle type of hydraulic amplifier; in turn, the flapper of the latter would be displaced by a force motor, the excitation of the coil of which would be controlled by a transistor power amplifier. Finally, the first stage would be a transistor or IC signal amplifier, driven by the analog error voltage from the digital-analog converter and driving the input to the power amplifier.

13.3.6   Development of the schematic diagram

Bearing in mind the comments outlined in Section 13.2, a schematic diagram can be constructed for the system as described so far: each system element selected can be represented by either a block or a pictorial symbol, as appropriate, and all interconnections (electrical, hydraulic, mechanical, etc.) can be incorporated to show how the elements are interlinked. In a real-world situation, terminal identifiers would be added to the diagram. Figure 13.1 gives an indication of the type of diagram (without identifiers) which would result. In a practical situation, waveforms, voltage and current levels, hydraulic supply pressures, etc. might also be added to the diagram.

Not shown in Fig. 13.1 are means for modifying and tuning the dynamic behaviour of the complete system. Typically, an active R-C network might be incorporated into the IC preamplifier; alternatively, a tachogenerator might be coupled mechanically to the leadscrew and its output combined with the error voltage (from the digital-analog converter), using the IC preamplifier in a signal combining configuration. Fine tuning would typically be implemented by adjusting the R and/or C values in the R-C filter or, in the alternative arrangement, by adjusting a potentiometer to manipulate the proportion of tachogenerator feedback voltage combined with the position error voltage. Having created a schematic diagram, the control engineer would then proceed through development stages 5 to 8 as listed in Section 13.3.

The particular example chosen represents a relatively complex and sophisticated type of control system. It was selected because it demonstrates a combination of a wide range of types of system element. Of course, it will be appreciated that the majority of control systems are relatively simple and have a less demanding performance specification than the example which has been presented.

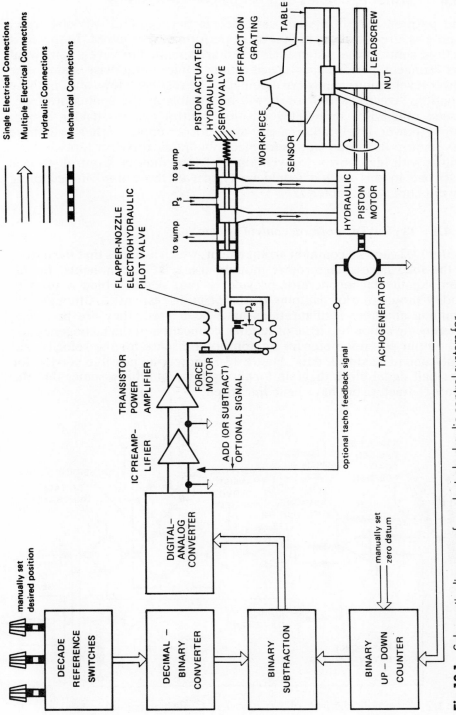

Fig 13.1   Schematic diagram of an electrohydraulic control system for precision positioning of a machine tool workpiece

13.4 Further examples of position control systems

The position control system described in Section 13.3 was elaborate, mainly as a result of the high level of performance required. In this next section, some of the systems described are assumed to have a somewhat less stringent performance specification. Unless otherwise stated, the power level requirements are assumed to be relatively low, and, in these situations, electric servomotors will be suitable as the final control elements: otherwise comparable position control systems having a high output power level requirement may, for example, either use these servomotors as torque motors for electrohydraulic drives or replace these motors with high power 'conventional' electric motors. Control systems described in this section would be typical of those used in recording instruments: see Chapter 7.

13.4.1 Typical DC position control system

Figure 13.2 shows a common arrangement, which assumes that the nature of the load necessitates rotary motion, using DC components. In this example, the tachogenerator provides a rate signal, which is used to modify the degree of damping of the position response. Although two summing amplifier configurations have been used, they are providing differencing action by virtue of suitable connection of the tachogenerator and output potentiometer for an appropriate sense for the velocity and displacement feedback data. Figure 13.3 shows one possible version for the small signal-block diagram for the system, and this assumes that the motor characteristics have been linearised.

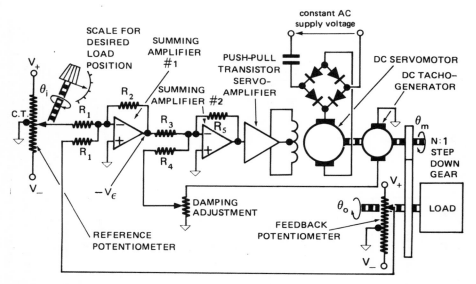

Fig 13.2 Schematic diagram of a low-power DC position control system

As an alternative to the tachogenerator, for providing damping action, an active R-C compensation network could be incorporated into the operational amplifier networks. Armature control of the servomotor would be a potential alternative to the field control arrangement shown.

13.4.2 Typical AC-carrier position control systems

Figure 13.4 shows an AC-carrier counterpart to the DC servosystem of Fig. 13.2. The AC carrier system will be different from the DC system in the following respects:

- there will be no discontinuity in the measuring system, because the synchro law is re-entrant
- the sychro system imposes a sine law upon the V_ε vs ε characteristic
- the servomotor torque vs speed characteristic synthesises an additional component of viscous friction, at the motor shaft.

The block diagram for this system will be similar in form to Fig. 13.3, except that the displacement transducer sensitivity will be dependent upon

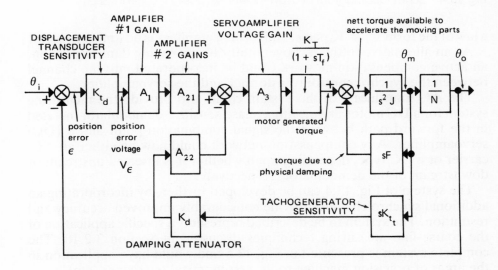

K_T	=	motor torque constant, in Nm/V
T_f	=	motor field time constant, in seconds
J	=	total polar moment of inertia of the moving parts, referred mathematically to the motor shaft
F	=	total viscous friction coefficient of the moving parts, referred mathematically to the motor shaft

Fig 13.3 Small-signal block diagram for the DC position control system depicted schematically in Fig13.2

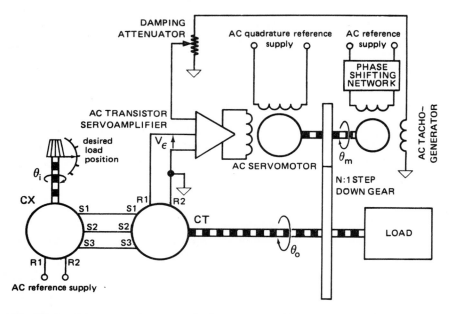

Fig 13.4   Schematic diagram of a low-power AC position control system

a law of the form $V_\varepsilon = K_{t_d} \sin \varepsilon$, and the first amplifier is no longer present.

As an alternative to the tachogenerator for providing damping action, an active compensation network could be incorporated into the channel between the synchro control transformer and the servoamplifier.

The AC transducers of this system could also provide the measuring system for a DC motor drive, with a phase-sensitive demodulator inserted in the forward path between the signal combination point and a (DC) servoamplifier. Any compensation network could now be either an AC-carrier or a DC type, depending upon whether it is inserted upstream or downstream of the demodulator, respectively.

The system of Fig. 13.4 can be developed further, by incorporating an additional synchro chain to provide considerably improved accuracy and resolution. The system to be described represents a specific application of the coarse-fine measuring techniques discussed in Section 3.2.10. The complete system is presented in Fig. 13.5 and would have application in the areas of precision machine tools, astronomical telescopes, etc.

Figure 13.6 shows the characteristics of the coarse and fine channel error voltages, plotted to a base of error $\varepsilon (= \theta_i - \theta_o)$. For ease of representation, a much smaller value has been chosen for the gear ratio N than would be used in a practical system. Also shown is one possible characteristic for the error voltage selection, although several alternatives are possible. This characteristic will emerge in the displacement transducer sensitivity block of any large-signal block diagram for the system, indicating that this will be a highly nonlinear system.

Figure 13.7 shows another coarse-fine system, in which the fine feed-

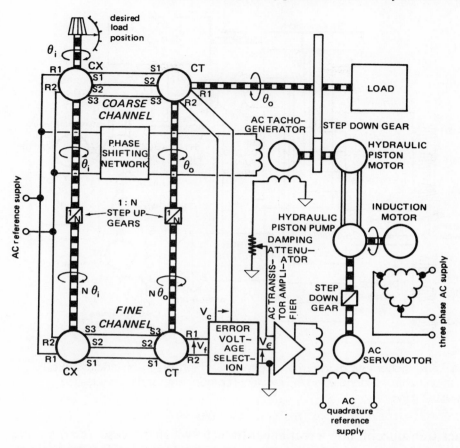

Fig 13.5   Schematic diagram of a high-precision electrohydraulic position control system, using AC electrical components and a coarse-fine measuring system

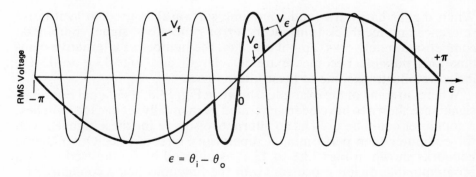

Fig 13.6   Error voltage characteristics for the coarse and fine measurement channels of the system of Fig 13.5. In practice, there would be many more cycles of the fine channel characteristic than have been shown here

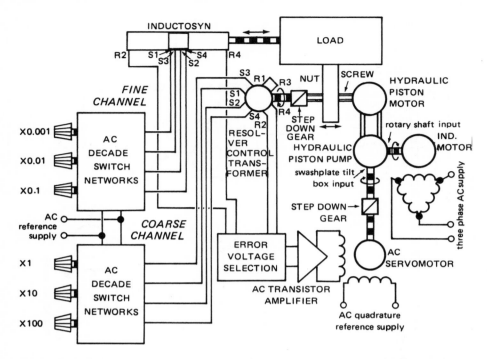

Fig 13.7 Schematic diagram of an alternative high-precision electrohydraulic position control system, using AC electrical components and a coarse-fine measuring system

back transducer is now a rectilinear Inductosyn. In this case, the reference voltages are assumed to be synthesised and selected by decade switches, using one of the types of network discussed in Section 3.3.3.

13.4.3 Typical numerical position control systems

When digital hardware (and sometime software) is used to synthesise reference and feedback transducers, error generators, signal combiners, compensation networks, amplifiers, etc., the number of alternative combinations possible becomes extremely large, and Figs 13.8 to 13.11, inclusive, illustrate some of these alternatives.

Where, in some of the examples shown in Fig. 13.8, the digital reference signals are shown to have been generated by manually set decade switches, a computer could be used as an alternative source for these signals, with the computer then performing a supervisory type of role. Other digital networks shown in Figs 13.8 to 13.11 could well be synthesised using a programmable digital processor, with the functions being configured by means of appropriate program statements.

In those configurations in Fig. 13.11 in which only sign of error and null error data are provided, the control logic must be configured so that the

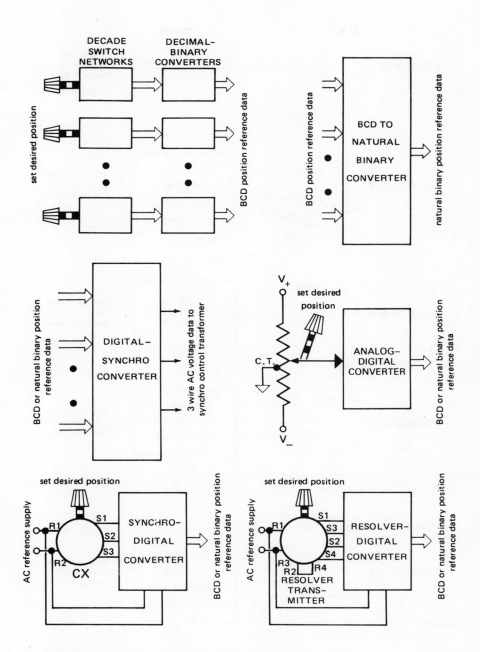

Fig 13.8 Some alternative numerical reference channels

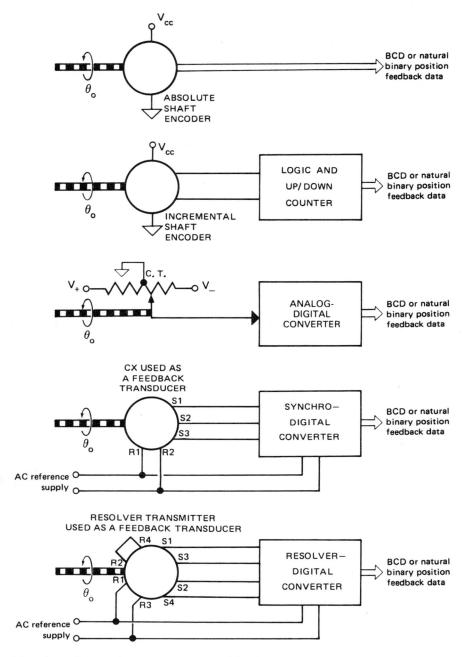

Fig 13.9   Some alternative numerical feedback channels

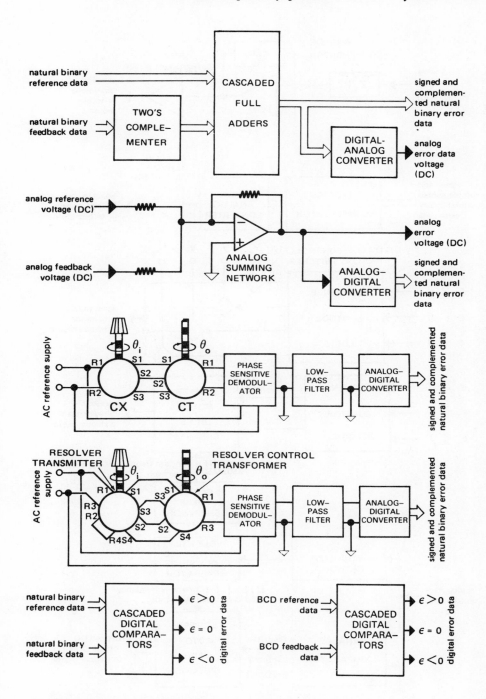

Fig 13.10 Some alternative numerical error data channels

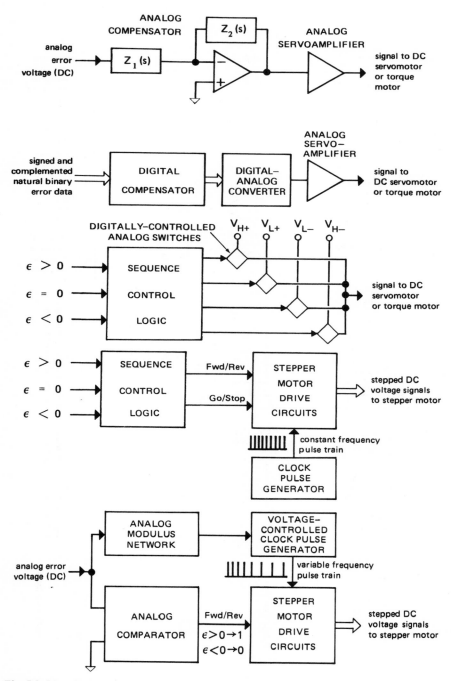

Fig 13.11 Some alternative error data processors

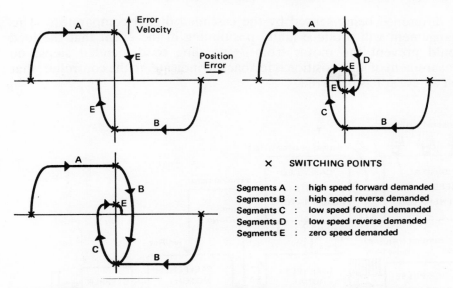

Fig 13.12   Some alternative numerical position control strategies using sequential selection of speed and direction

position error is driven towards zero at sequentially stepped speeds. Representative 'phase portraits' of error velocity vs position error are shown in Fig. 13.12, to illustrate some possible positioning routines: these routines would only be usable provided that the position reference value is stationary during the positioning sequences. This distinction results in these systems sometimes being called 'point-to-point positioning systems', in contrast to 'continuous path positioning systems'.

As an alternative to switching the motor to drive at various fixed speeds, it would be possible to use the signals to engage and disengage clutches in a clutch-gear system, driven by a single-speed motor, of the type shown in Fig. 12.20. Obviously, if the drive is brought to rest from a fairly low speed condition, the ultimate position error will be small but the positioning time may be large. If the sequence results in final stopping always from the same direction, irrespective of the sign of the initial position error, then the ultimate position error should be more consistent than would otherwise be the case, especially when the values of the friction and inertia of the moving parts are consistent. When the initial value of position error is small, the sequence controller should be arranged so that full speed is always attained during the positioning cycle, in order to achieve consistency of stopping.

Figure 13.13 shows one possible system in which a stepper motor can be used for open loop position control, and this is one of the few cases where open loop control of position is practicable. The motor is driven in such a manner that the current output count from the bidirectional counter is always driven towards that value represented by the desired position data, the difference being sensed by the cascaded digital comparators. This arrangement will produce faulty positioning in the event that the load should prevent the motor from responding to demanded steps: no measurement of load position is fed back to indicate, to the controller, that such an event has occurred.

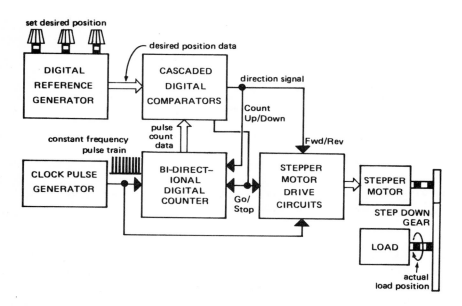

Fig 13.13   A typical scheme for open loop position control using a stepper motor

13.5   Examples of speed control systems

In Sections 11.4.1 and 11.4.2, commercially available speed controllers for conventional DC and AC motors were described. In the event that these should be unsuitable for a particular application, it becomes necessary to custom design a speed controller. Whilst almost any closed loop speed control system will still function, after a fashion, as an open loop speed control system, if the feedback is disconnected, accurate speed control will rarely eventuate in this situation.

At low output power levels, servomotors may be used to provide drives for instrument types of application. At higher levels of output power, these servomotors can be used as torque motors to actuate hydraulic servovalves or pumps which, in turn, would manipulate hydraulic rotary actuators, cylinders, motors, etc. Alternatively, speed controllers using the same principles as commercial types may be custom designed, to operate with electric motors. A few examples of low power configurations will be described, in Sections 13.5.1, 13.5.2, and 13.5.3.

13.5.1 Speed control of a typical small DC motor drive

Figure 13.14 shows a typical DC instrument velocity servosystem. It will be seen to be similar to the position servosystem of Fig. 13.2, except that the position feedback of the latter now is absent. The summing amplifier provides a differencing action, by virtue of suitable connection of the tachogenerator for appropriate sense for the velocity feedback data. In the absence of any suitable (minor) feedback transducer for compensation use, compensation would normally be provided by an active or passive R-C network in the forward path. Armature control of the servomotor would be a potential alternative to the field control arrangement shown.

For the selection of preset speed references, a potentiometric reference network of the type described in Section 3.3.1 would be used to replace the reference potentiometer shown in Fig. 13.14. The supply to the reference transducer need only be unipolar if the drive is to be unidirectional.

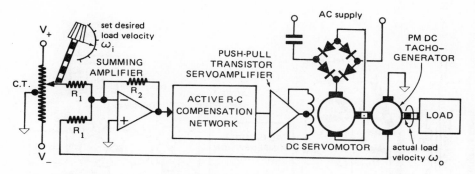

Fig 13.14 Schematic diagram of a low-power DC velocity control system

Figure 13.15 shows one possible version of the small-signal block diagram for the system of Fig. 13.14.

13.5.2 Speed control of a typical small AC motor drive

Figure 13.16 shows a typical AC instrument velocity servosystem. It will be seen to be very similar to its DC counterpart of Fig. 13.14. The transformer connected to the reference potentiometer has been provided in order to establish a supply symmetrically balanced about signal common, on the assumption that the motor drive is to be bidirectional: this transformer is redundant for unidirectional drives. An alternative, and potentially superior, reference transducer would be an auto-transformer, with either a continuously variable wiper output or switched output tappings. The summing amplifier provides a differencing action, by virtue of suitable connection of the tachogenerator for appropriate sense for the velocity feedback data. In the absence of any suitable (minor) feedback transducer for compensation use, compensation would normally be provided by an

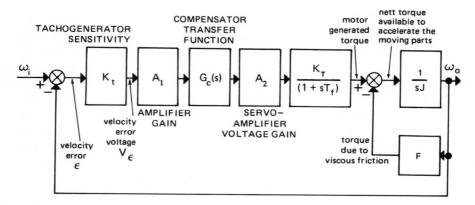

Fig 13.15　Small-signal block diagram for the DC velocity control system depicted schematically in Fig 13.14

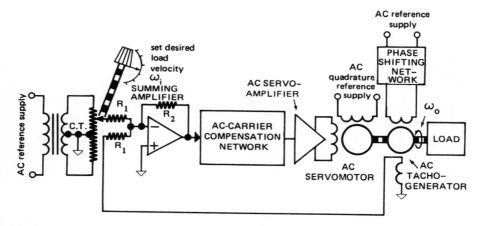

Fig 13.16　Schematic diagram of a low-power AC velocity control system

AC-carrier compensation network in the forward path. The small-signal block diagram for this system would resemble that shown in Fig. 13.15.

13.5.3　Speed control of stepper motors

Provided that a stepper motor can develop sufficient torque to drive the load for which it has been chosen, its speed can be controlled accurately without the need for velocity feedback. Because the stepping rate will be proportional to the frequency of the pulse train supplied to the drive circuits, a speed controller for this type of motor will be required to establish a precisely varied pulse frequency. In Section 11.4.3, some types of stepper motor speed controllers were described: in these, the pulse train was developed by a voltage-controlled oscillator, with the applied DC control voltage determining the pulse frequency. An alternative technique is to use digital means to manipulate the frequency, and an example is shown in Fig. 13.17. The counter could be arranged so that the frequency of the pulse train supplied to the stepper motor drive circuits is equal to the clock pulse frequency divided by the output word from the digital code converter, so that the latter will need to be representative of the reciprocal of the desired load speed. The drive could be made bidirectional by causing a forward/reverse switch on the set point station to reverse the sense in which the DC output voltages are circulated around the motor windings.

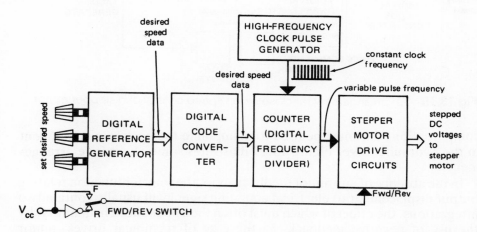

Fig 13.17 An arrangement for open loop speed control of a stepper motor

When closed loop control using numerical means is required, this may be undertaken using a scheme like that of Fig. 13.18. The algorithm of the digital compensator should include a synthesis of integration, so that the value of pulse frequency necessary to drive the load at the desired speed can be sustained even when the speed error has been reduced to zero. The digital frequency multiplication is best implemented by reciprocating the data from the compensator and then using a pulse counter as a frequency divider. The system shown is best used for unidirectional drives, although conceivably it could be adapted for bidirectional action, possibly by using the forward/reverse arrangement in the open loop system of Fig. 13.17.

13.6 Examples of electrohydraulic drives

In those servosystems in which the drive is required to develop high power levels, it becomes necessary to use either large conventional electric motors or hydraulic drives. In the latter case, the error signal of the control system will usually be electrical and, after suitable amplification and

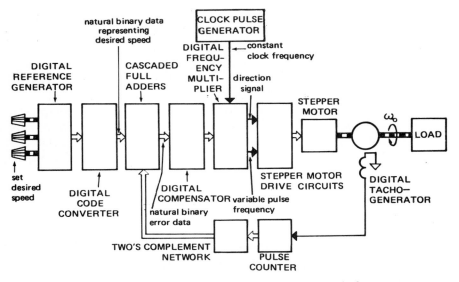

Fig 13.18   An arrangement for closed loop speed control of a stepper motor

compensation, this must be used to manipulate the upstream component in the hydraulic drive. Figure 13.19 illustrates some alternative arrangements.

In the absence of any minor feedback loop, the transfer function relating output displacement to electrical actuating signal will usually contain two integrations, the effect of which must often then be reduced in quantity, by the use of negative feedback. In the case of rectilinear drives, minor feedback may be implemented using mechanical linkages, connected from the output shaft back to the servovalve, which may incorporate an inner sleeve. Minor feedback may also be effected electrically, using transducers which sense the pressure of the manipulated hydraulic supply to the motor, cylinder, or actuator and feed back into the amplifier which is developing the electrical actuating signal. Feedback will also have a linearising effect upon the static characteristic and will modify the speed of response of the components around which it has been connected. The servomotor shown in Fig. 13.19 may be of the DC, AC, or stepper type, which may be used as a torque motor.

13.7   Examples of process control systems

In Section 11.3, it was explained that general-purpose process controllers could be configured as feedback, cascade, feedforward, and ratio controllers, each having an appropriate control law. In this section, a few examples will be described, to demonstrate their application. Figure 13.20 shows some typical symbols used in schematic diagrams for process loops.

In many industrial diagrams, distinctive symbols will be used, to distinguish between electrical, hydraulic, and pneumatic interconnections. In

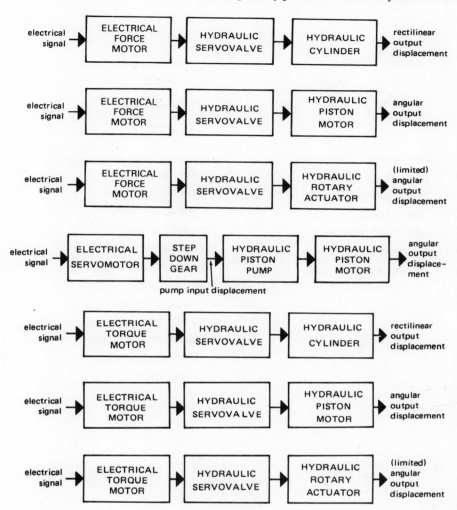

Fig 13.19 Some alternative arrangements of hardware for electrohydraulic drives

process mimic diagrams, it is normal practice to allocate different colours to process loops controlling different properties of the plant or different chemicals and materials within the plant.

13.7.1 Typical process loop using a feedback controller

A very simple example of the use of a feedback controller would be for the control of the level of liquid in a storage tank, by manipulating the inflow rate, with outflow rate being a wild variable. Such a system is illustrated in Fig. 13.21.

TRANSMITTERS

TT	temperature transmitter
PT	pressure transmitter
FT	flow transmitter
LT	level transmitter
DT	density transmitter
pHT	pH transmitter
CT	composition transmitter

CONTROLLERS

xC	controller, with no indication or recording
xIC	indicating controller
xRC	recording controller
TxC	temperature controller
PxC	pressure controller
FxC	flow controller
LxC	level controller
DxC	density controller
pHxC	pH controller
CxC	composition controller
FFC	feedforward controller
RC	ratio controller

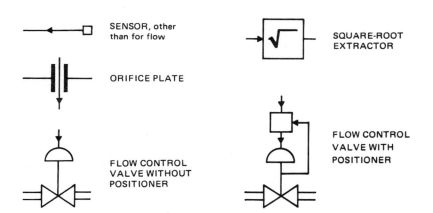

Fig 13.20   Some typical symbols in schematic diagrams for process control loops. x represents an upper-case letter, to be specified according to the application

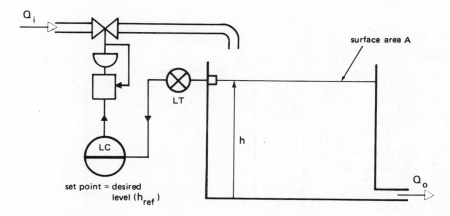

Fig 13.21 Schematic diagram of a liquid level control system using a feedback process controller

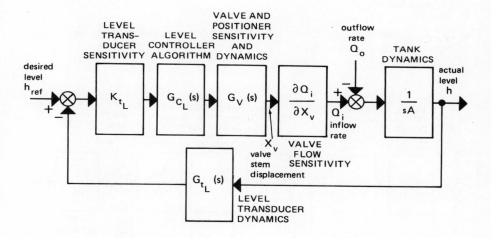

h_{ref}, h, Q_i, Q_o are increments in level and flow.
X_v is increment in valve stem displacement.

Fig 13.22 Small-signal block diagram for the liquid level control system depicted in Fig 13.21

A small-signal block diagram for this type of configuration would take the form shown in Fig. 13.22.

13.7.2 Typical use of a cascade controller

In the previous example, the nonlinearity of the control valve flow characteristic and possible slowness in the valve speed of response can

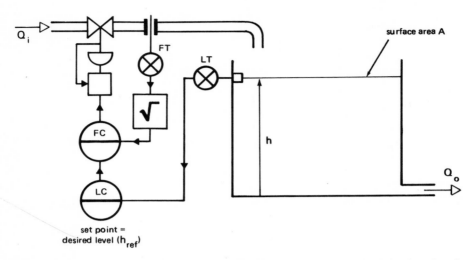

Fig 13.23   Schematic diagram of a liquid level control system using feedback and cascade process controllers

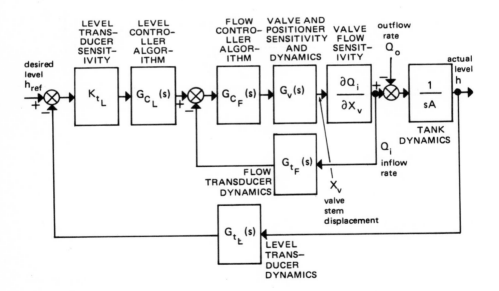

Fig 13.24   Small-signal block diagram for the liquid level control system depicted in Fig13.23

both be alleviated by the introduction of a minor feedback loop, in which the inflow rate is measured and fed to a flow controller. Figure 13.23 illustrates this modified system. In this arrangement, the level controller performs the role of a feedback controller and its output becomes the (remote) set point for the flow controller, which acts as a cascade controller. Figure 13.24 illustrates the small-signal block diagram, and it can be seen that the level loop is the outermost (major) loop, with greater authority than the flow loop, which is the inner, minor, loop.

13.7.3 Typical use of a feedforward controller

In the example of Fig. 13.21, the feedback controller has to attempt to hold the liquid level at the set point value, despite changes in the wild variable, the outflow rate Q_o. If, in a practical case, the controller cannot achieve the desired accuracy, then the situation can be relieved by instrumenting the wild variable and incorporating a feedforward controller, as shown in Fig. 13.25.

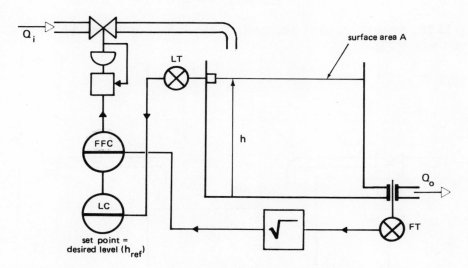

Fig 13.25 Schematic diagram of a liquid level control system using feedback and feedforward process controllers

Reference to the small-signal block diagram of Fig. 13.26 shows that the new controller does not provide additional feedback action, because it does not create a new closed loop, but it enables compensation to be effected for fluctuations in Q_o.

With the plant under discussion, both feedforward and cascade controllers could conceivably be incorporated with the feedback controller, in the one system.

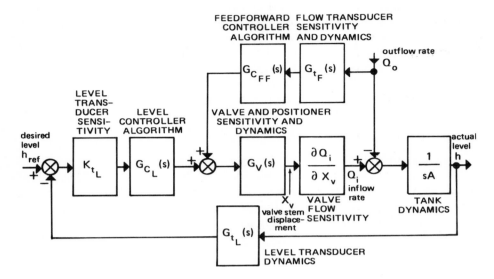

Fig 13.26   Small signal block diagram for the liquid level control system depicted in Fig 13.25

Ideally $G_{t_F}(s) \cdot G_{c_{FF}}(s) \cdot G_v(s) \cdot \dfrac{\partial Q_i}{\partial X_v} = 1$

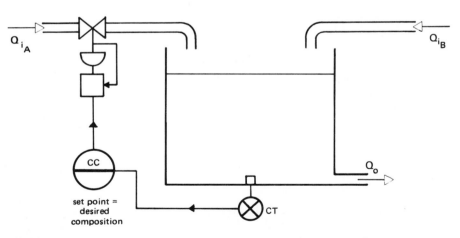

Fig 13.27   Schematic diagram of a liquid composition control system using a composition transducer and feedback process controller

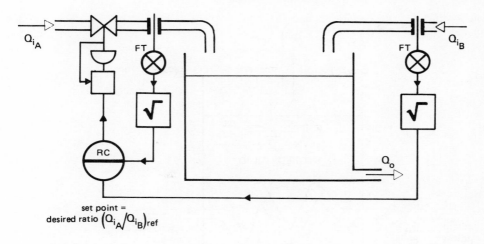

Fig 13.28 Schematic diagram of a liquid composition control system using a ratio process controller

13.7.4 Typical process loop using a ratio controller

In this example, inflows of two different liquids are mixed in a hold-up vessel and the flow rate of one is to be manipulated in order to maintain the composition of the mixture in the vessel at a desired consistency. The inflow rate of the other liquid is not controlled by this process and therefore is regarded as being wild.

One approach is to measure the composition of the mixture with a suitable composition transducer and to incorporate this with a composition controller, as shown in Fig. 13.27.

If a suitable composition transducer cannot be obtained, for whatever reason, then ratio control can be used as a possible alternative, with the two inflow rates being measured and their (controlled) ratio being used to infer the current composition of the mixture: such a scheme is shown in Fig. 13.28.

It will be seen that any long-term error in the actual value achieved for the ratio may result in a significant error in the composition of the mixture, and there is no instrumentation to indicate directly that this situation has arisen.

13.7.5 Representative temperature control system

Figure 13.29 shows a schematic arrangement of the system components to achieve the control of temperature of liquid held in a vessel which encloses a heat exchanger.

The control valve is used to manipulate the flowrate of steam entering the heat exchanger, such that the rate at which heat is transferred to the

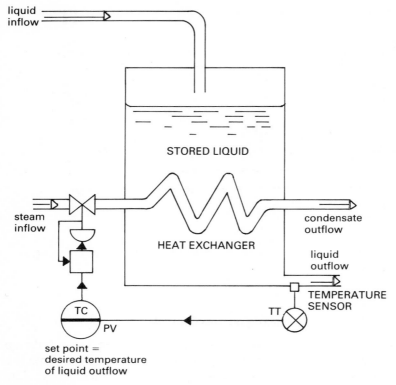

liquid
inflow

STORED LIQUID

steam
inflow

condensate
outflow

HEAT EXCHANGER

liquid
outflow

TEMPERATURE
SENSOR

TC

PV

TT

set point =
desired temperature
of liquid outflow

Fig 13.29 Schematic diagram of a simple temperature loop controlling a storage vessel containing a heat exchanger

liquid in the vessel is of such a magnitude as to cause the actual temperature of the liquid outflow to match the desired temperature established as the set point of the temperature controller.

The temperature sensor could be one of the many alternatives described in Section 5.3. The temperature transmitter will be required to condition the signal generated by the sensor and convert it to lie within the (probably standard) PV input signal range of the temperature controller.

The performance of this control loop could easily be degraded by large variations in one or more of the following parameters:

- the temperature of the steam inflow

- the temperature of the liquid inflow

- the volume of liquid held up in the vessel

- the thermal properties of the liquid in the vessel.

If the temperature loop is unable, acting alone, to hold the temperature error excursions within acceptable limits, then it must be assisted in its task: normally, this would be achieved by transducing the offending wild variable and using the signal thereby generated to effect feedforward action, by the insertion of a feedforward controller between the temperature controller and the valve positioner.

Cascaded steam flow control could also be introduced by transducing the steam flowrate and inserting a cascaded (flow) controller to drive the input to the valve positioner. This would help to nullify the effect of nonlinearity in the valve installed characteristic, any sluggishness in the valve action, and any large wild variations in the steam supply pressure.

Needless to say, the small-signal block diagram for any of these more complex configurations will itself be complex, resulting in increasingly complicated analysis and tuning procedures.

13.7.6   Representative pressure control system

Figure 13.30 is a symbolic representation of a coal-fired boiler, showing the system for controlling the steam pressure developed in the boiler.

The rotating ball mill pulverises coal which falls into the top of the mill from a hopper, the cast-iron balls crushing the lumps of coal as the mill rotates. The coal dust thereby created is blown, by the primary air fan, into the furnace where it is ignited by a flame sustained by a fuel oil jet. The heat generated by the combustion converts the boiler water into superheated steam, which is drawn off from the top of the boiler, to be used (say) for powering a steam turbine.

The rate of heat generation is determined by the rate at which coal fuel is blown into the boiler, and this in turn depends upon the displacement of air by the primary air fan. The pitch of each of the fan blades (the set of which is driven at constant speed by an induction motor) is manipulated by means of a vane ring, the displacement of which is determined by the action of the vane ring actuator. The actuator, displacement transducer and actuator controller form a closed loop rectilinear displacement control system—in other words, a rectilinear servosystem.

The reference variable of the actuator controller is desired vane ring displacement, and the corresponding reference signal is generated by the output from the pressure controller, the process variable of which is steam outflow pressure. The performance of the pressure control loop would typically be enhanced by the addition of a feedforward controller inserted between the pressure controller and the actuator controller. Feedforward action would be provided by sensing and feeding forward steam outflow rate, using (say) an orifice plate, flow transmitter and square-root extractor. In this manner, excursions in the steam outflow pressure error can be minimised despite possibly large and sudden changes in steam outflow demanded by the turbine.

13.7.7   Representative flow control system

As an example of a system to control fluid flowrate, consider the require-

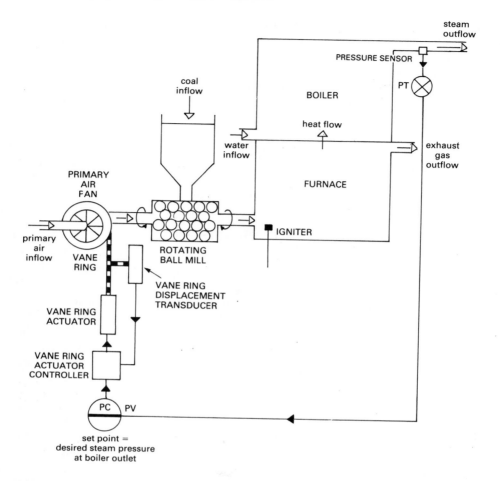

Fig 13.30   Schematic diagram of a pressure loop controlling the steam pressure developed in a boiler

ment for controlling the flow of water from a storage tank to a boiler 'drum' (or 'header tank'), which could be used, for example, to supply the feedwater for the boiler in the previous example.

Assuming that the drum is situated many metres above the level of the storage tank, the water needs to be lifted by a pump. Flowrate could be manipulated by varying the speed of the motor driving the pump, by varying the throttling action of a control valve with positioner (the pump now being driven at constant speed), or by varying the setting of a torque-converting fluid coupling placed between a constant-speed motor and the now variable-speed pump. Figure 13.31 shows how these configurations could be organised.

In practice, it would be necessary to establish a liquid level loop around each of these flow loops, in order to maintain the water in the drum at a

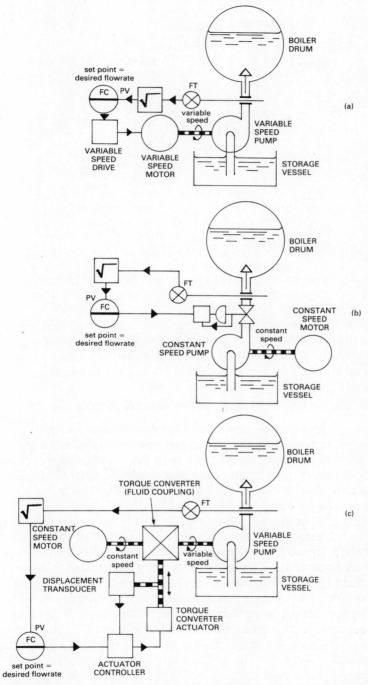

Fig 13.31 Schematic diagram for three alternative configurations for achieving flowrate control of feedwater to a boiler drum, using (a) a variable speed pump, (b) a flow control valve, (c) a torque converter

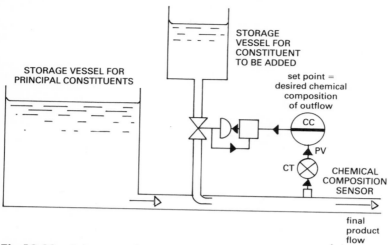

Fig 13.32 Schematic diagram of a typical chemical composition control loop

desirable level. The output of the level loop would become the remote set point input of the flow controller, which now functions as a cascade controller.

It is worth recording that techniques such as those described here can be used to control the flowrates of any fluids, such as liquids, vapours, gases and steam. Clearly, the choice of flowrate sensor and final control element will be influenced strongly by the type of fluid being manipulated.

13.7.8   Representative chemical composition control system

Many alternative configurations could be described in order to represent chemical composition control systems. As just one example, assume that the product from a manufacturing process is in the form of liquid being produced continuously (that is, rather than as a result of a 'batch' process). Suppose that a constituent component also is a liquid, which has to be metred via a flow control valve, as shown in Fig. 13.32.

The composition of the final product may possibly be measured using many alternative techniques, depending upon the nature of the product and the availability of suitable on-stream analysers. For example, composition might be inferred by measuring thermal conductivity, colour, opacity, pH, or it might be measured more directly using chromatography and/or spectrometry. The measurement may generate continuous (PV) data, or it may generate only sampled data if the analyser needs to process samples of the product stream. Chemical reaction may be occurring between the point at which the new constituent is added and the site of the chemical composition sensor. The calibration of sensor output signal versus actual chemical composition may be highly nonlinear, in which case the composition controller could be required to apply a linearisation algorithm to its PV data. Additionally, the relationship between change in chemical composition and proportion of added constituent may also be highly nonlinear.

Bibliography

General control system theory

Anand, Davinder K. (1984) *Introduction to control systems*. 2nd ed. Pergamon Press, Oxford, New York.

Atkinson, Peter (1972) *Feedback control theory for engineers*. 2nd ed. Heinemann, London.

Auslander, David M. *et al.* (1974) *Introducing systems and control*. McGraw-Hill, New York.

Bateson, Robert (1980) *Introduction to control system technology*. 2nd ed. Merrill, Columbus, Ohio.

Bissell, Chris C. (1988) *Control engineering*. Van Nostrand Reinhold, New York.

Buckley, Ruth Victoria (1976) *Control engineering: theory, worked examples and problems*. Macmillan, London, New York.

Chestnut, Harold and Mayer, R. W. (1979) reprint *Servomechanisms and regulating system design*. Wiley, New York.

Cruz, Jose Bejar. (1972) *Feedback systems*. McGraw-Hill, New York.

D'Azzo, John Joachim and Houpis, C. H. (1981) *Linear control system analysis and design: conventional and modern*. 2nd ed. McGraw-Hill, New York.

Distefano, Joseph J. *et al.* (1967) *Schaum's outline of theory and problems of feedback and control systems*. McGraw-Hill, New York.

Dorf, Richard C. (1980) *Modern control systems*, 3rd ed. Addison-Wesley, Reading, Mass.

Emanuel, Pericles. (1979) *Introduction to feedback control systems*. Mc-Graw-Hill, New York.

Eveleigh, Virgil W. (1971) *Introduction to control systems design*. McGraw-Hill, New York.

Fortmann, Thomas E. (1977) *An introduction to linear control systems*. M. Dekker, New York.

Franklin, Gene F. *et al.* (1986) *Feedback control of dynamic systems*. Addison-Wesley, Reading, Mass.

Furuta, K. *et al. State variable methods in automatic control*. John Wiley, Chichester, U.K.

Healey, Martin. (1975) *Principles of automatic control*. 3rd ed. English Universities Press, London.

Hotstetter, Gene H. *et al.* (1982) *Design of feedback control systems*. Holt, Rinehart and Winston, New York.

Humphrey, William M. (1973) *Introduction to servomechanism system design.* Prentice-Hall, Englewood Cliffs, N.J.

Kuo, Benjamin. (1987) *Automatic control systems.* 5th ed. Prentice-Hall, Englewood Cliffs, N.J.

McDonald, Anthony C. and Lowe, H. (1981) *Feedback and control systems.* Reston Pub. Co., Reston, Va.

MacFarlane, Alistair G. J. (1979) *Frequency-response methods in control systems.* IEEE, Piscataway, N.J.

Miller, Richard W. (1977) *Servomechanisms: devices and fundamentals.* Reston Pub. Co., Reston, Va.

Morris, Noel Malcolm (1974) *Control engineering.* 2nd ed. McGraw-Hill, London, New York.

Nagrath, I. J. and Gopal, M. (1982) *Control systems engineering.* 2nd ed. Wiley, New York.

Ogata, Katsuhiko. (1970) *Modern control engineering.* Prentice-Hall, Englewood Cliffs, N.J.

Phelan, Richard M. (1977) *Automatic control systems.* Cornell University Press, Ithaca, N.Y.

Poucher, George Howard (1974) *Introduction to control techniques.* Pitman, London.

Raven, Francis Harvey. (1978) *Automatic control engineering.* 3rd ed. McGraw-Hill, New York.

Richards, R. J. (1979) *An introduction to dynamics and control.* Longman, London.

Rubin, Olis. (1986) *The design of automatic control systems.* Artech House, Norwood, Ma.

Sante, Daniel P. (1980) *Automatic control system technology.* Prentice-Hall, Englewood Cliffs, N.J.

Schwarzenbach, J. and Gill, K. F. (1984) *System modelling and control.* 2nd ed. Edward Arnold, London.

Shinners, Stanley Marvin. (1978) *Modern control system theory and application.* 2nd ed. Addison-Wesley, Reading, Mass.

Sinha, Naresh Kumar. (1986) *Control systems.* Holt, Rinehart and Winston, New York.

Truxal, John G. (1972) *Introductory system engineering.* McGraw-Hill, New York.

Weyrick, Robert C. (1975) *Fundamentals of automatic control.* McGraw-Hill, New York.

Wilson, Derek Robert. (1970) *Modern practice in servo design.* Pergamon Press, Oxford, New York.

Wolsey, William Harry. (1975) *Basic principles of automatic control: with special reference to heating and air conditioning systems.* Hutchinson, London.

Process control systems

Cheremisinoff, Paul N. (1981) *Automatic process control.* Ann Arbor Science, Ann Arbor, Michigan.

Instrument Society of America. (1988) *Introduction to process control: principles and applications.* The Society, Pittsburgh.

Johnson, Curtis D. (1988) *Process control instrumentation technology.* 3rd ed. Wiley, New York.

Murrill, Paul W. (1988) *Application concepts of process control.* Instrument Society of America, Pittsburgh.

Shinskey, F. G. (1988) *Process–control systems: application, design and tuning.* 3rd ed. McGraw-Hill, New York.

Weber, Thomas W. (1973) *An introduction to process dynamics and control.* Wiley, New York.

Digital control systems

Blaschke, W. S. and McGill, J. (1976) *The control of industrial processes by digital techniques: the organisation, design and construction of digital control systems.* Elsevier, Amsterdam.

Franklin, Gene F. *et al.* (1986) *Digital control of dynamic systems.* Addison-Wesley, Reading, Mass.

Kuo, Benjamin C. (1980) *Digital control systems.* Holt, Rinehart and Winston, New York.

Moorehead, Jack. (1980) *Numerical control: vol. 1. Fundamentals.* Society of Manufacturing Engineers, Dearborn, Michigan.

Moorehead, Jack. (1980) *Numerical control: vol. 2. Application.* Society of Manufacturing Engineers, Dearborn, Michigan.

Ogata, Katsuhiko. (1987) *Discrete-time control systems.* Prentice-Hall, Englewood Cliffs, N.J.

Signal theory and processing

Arbel, A. F. (1984) *Analog signal processing and instrumentation.* Cambridge University Press, Cambridge, Eng.

De Coulon, Frederic. (1986) *Signal theory and processing.* Artech House, Norwood, Ma.

Picinbono, Bernard. (1988) *Principles of signals and systems: deterministic signals.* Artech House, Norwood, Ma.

Rabiner, Lawrence R. and Rader, C. N. (1972) *Digital signal processing.* IEEE, Piscataway, N.J.

Van der Ziel, Albert. (1976) *Noise in measurements.* Wiley Interscience, New York.

Measurement, transducers and instrumentation

Adams, Leslie Frank. (1975) *Engineering measurements and instrumentation.* Hodder and Stoughton, London.

Andrew, William G. (1979) *Applied instrumentation in the process industries.* 2nd ed. Gulf Pub. Co., Houston.

Baiulescu, George and Ilic, V. A. (1975) *Stationary phases in gas chromatography.* Pergamon Press, Oxford.

Bannister, Brian Roy and Whitehead, D. G. (1986) *Transducers and interfacing: principles and techniques.* Van Nostrand Reinhold, Wokingham, U.K.

Beckwith, Thomas G. and Buck, N. L. (1982) *Mechanical measurements.* 3rd rev. ed. Addison-Wesley, Reading, Mass.

Bell and Howell Company, CEC/Instruments Division. (1974) *Pressure transducer handbook.* The Company, Pasadena, Calif.

Bentley, John P. (1983) *Principles of measurement systems.* Longman, London, New York.

Borer, John R. (1985) *Instrumentation and control for the process industries.* Elsevier Scientific, London, New York.

Bristow, Paul Anthony. (1976) *Liquid chromatography in practice.* Hetp, Handforth.

Cheremisinoff, Nicholas P. (1981) *Process level instrumentation and control.* M. Dekker, New York.

Clayton, C. G. (1972) *Modern developments in flow measurement.* Peregrinus, London.

Cluley, John Charles. (1985) *Transducers for microprocessor systems.* Macmillan, London.

Conder, John R. and Yeung, C. L. (1979) *Physicochemical measurement by gas chromatography.* Wiley, Chichester, U.K.

Considine, D. M. (1981) reprint *Encyclopedia of instrumentation and control.* McGraw-Hill, New York.

Considine, D. M. (1985) *Process instruments and controls handbook.* 3rd rev. ed. McGraw-Hill, New York.

Considine, D. M. (1987) *Standard handbook of industrial automation.* Chapman and Hall, London.

Cooper, William David. (1985) *Electronic instrumentation and measurement techniques.* 3rd ed. Prentice-Hall, London.

Engineering Equipment Users Association (Great Britain). (1973) *Installation of instrumentation and process control systems.* Constable, London.

Fribance, Austin E. (1962) *Industrial instrumentation fundamentals.* McGraw-Hill, New York.

Hayward, A. T. J. (1979) *Flowmeters: a basic guide and source-book for users.* Macmillan, London.

Hearn, Edward John. (1971) *Strain gauges.* Merrow, Watford, U.K.

Herceg, Edward E. (1972) *Handbook of measurement and control: an authoritative treatise on the theory and application of the LVDT.* Schaevitz Engineering, Pennsauken, N.J.

Hougen, Joel O. (1979) *Measurements and control applications.* 2nd ed. Instrument Society of America, Pittsburgh.

Hunter, Richard Sewall. (1976) *The measurement of appearance.* Wiley, New York.

Instrumentation Society of America. (1987) *Advances in instrumentation: proceedings of the ISA/87 International Conference and Exhibition.* The Society, Pittsburgh.

Instrumentation Society of America (1988) *Instrumentation in the chemical and petroleum industries, vol. 20: proceedings of the chemical and petroleum industry division symposium, Tulsa, Oklahoma, 1988.* The Society, Pittsburgh.

Instrument Society of America. (1988) *Resource management through better instrumentation: technical papers . . . proceedings of the Pacific Cascade Instrumentation '88 Exhibition and Symposium, Eugene, Oregon, 1988.* The Society, Pittsburgh.

Johnson, Curtis D. (1982) *Process control instrumentation technology.* 2nd ed. Wiley, New York.

Jones, Ernest Beachcroft. (1985) *Instrument technology, vol. 1: mechanical measurements.* 4th ed. Newnes-Butterworth, London.

Jones, Ernest Beachcroft. (1985) *Instrument technology, vol. 2: measurement of temperature and chemical composition.* 4th ed. Newnes-Butterworth, London.

Jones, Ernest Beachcroft. (1986) *Instrument technology, vol. 5: automatic instruments and measuring systems.* 4th ed. Newnes-Butterworth, London.

Knapman, C. E. H. (1978–1980) *Developments in chromatography (2 vols).* Applied Science Pubs., London.

Kosow, Irving L. (1988) *Electronic instrumentation.* Wiley, New York.

Krohn, D. A. (1988) *Fiber optic sensors.* Instrument Society of America, Pittsburgh.

Kychakoff, G. (1988) *Electro-optic sensing and measurement: proceedings of the 6th International Congress on Applications of Laser and Electro-Optics (ICALEO '87).* Springer-Verlag, Berlin.

Lang, Tran Tien. (1987) *Electronics of measuring systems.* John Wiley, Chichester, U.K.

Levsen, Karsten. (1978) *Fundamental aspects of organic mass spectrometry.* Verlag Chemie, New York, Berlin.

Lion, Kurt Siegfried. (1975) *Elements of electrical and electronic instrumentation: an introductory textbook.* McGraw-Hill, New York.

Littlewood, A. B. (1970) *Gas chromatography: principles, techniques and applications.* 2nd ed. Academic Press, New York.

McFadden, William H. (1988) reprint. *Techniques of combined gas/chromatography/mass spectroscopy: applications on organic analysis.* Wiley, New York 1973. Kreiger, New York.

McLafferty, Fred Warren. (1980) *Interpretation of mass spectra.* University Sc. Books, Mill Valley, Ca.

Mansfield, P. H. (1973) *Electrical transducers for industrial measurement.* Butterworths, London.

Marcus, Abraham and Lenk, John D. (1971) *Measurements for technicians.* Prentice-Hall, Englewood Cliffs, N.J.

Middleditch, Brian S. (1979) *Practical mass spectrometry: a contemporary introduction.* Plenum Press, New York.

Mikes, O. (1979) *Laboratory handbook of chromatographic and allied methods.* Ellis Horwood, Chichester, Eng.

Morrison, Ralph. (1986) *Grounding and shielding techniques in instrumentation.* 3rd ed. Wiley-Interscience, New York.

Mylroi, M. G. and Calvert, G. (1984) *Measurement and instrumentation for control.* Peregrinus for the IEE, London.

National Semiconductor Corporation. (1977) *Data acquisition handbook.* The Corporation, Santa Clara, Cal.

National Semiconductor Corporation. (1977) *Pressure transducer handbook.* The Corporation, Santa Clara, Cal.

Neubert, Hermann K. P. (1975) *Instrument transducers: an introduction to their performance and design.* 2nd ed. Clarendon Press, Oxford.

Norton, Harry N. (1984) *Sensor selection guide.* Elsevier Sequoia, New York.

Oliver, Frank J. (1972) *Practical instrumentation transducers.* Pitman, London.

Parris, N. A. (1984) *Instrumental liquid chromatography: a practical manual on high-performance liquid chromatographic methods.* 2nd ed. Elsevier Scientific, Amsterdam.

Pattison, James Bulmer. (1973) *A programmed introduction to gas–liquid chromatography.* 2nd ed. Heyden, London.

Pryde, Andrew and Gilbert, M. T. (1979) *Applications of high performance liquid chromatography*. Chapman and Hall, London.

Rhodes, Thomas J. (1972) *Industrial instruments for measurement and control*. 2nd ed. McGraw-Hill, New York.

Seippel, Robert G. (1983) *Transducers, sensors and detectors*. Reston Pub. Co., Reston, Va.

Snell, Foster D. (1978) *Photometric and fluorometric methods of analysis: metals*. Wiley, New York.

Snell, Foster D. (1981) *Photometric and fluorometric methods of analysis: non-metals*. Wiley-Interscience, New York.

Snell, Foster D. and Snell, Cornelia T. (1967–1970) *Colorimetric methods of analysis: vols 4A-4AAA*. Van Nostrand, Reinhold, New York.

Snyder, Lloyd R. and Kirkland, J. J. (1979) *Introduction to modern liquid chromatography*. 2nd ed. Wiley, New York.

Sydenham, Peter Henry (1979) *Measuring instruments: tools of knowledge and control*. Peregrinus, Stevenage, Eng.

Sydenham, Peter Henry. (1985) *Transducers in measurement and control*. 3rd ed. A. Hilger, Bristol.

Usher, M. J. (1985) *Sensors and transducers*. Macmillan, London.

Window, A. L. and Holister, G. S. (1982) *Strain gauge technology*. Applied Science, London.

Woolvet, G. A. (1979) *Transducers in digital systems*. Rev. ed. Peregrinus for the IEE, Stevenage, U.K.

Zweig, Gunter and Sherma, Joseph. (1973) *CRC handbook of chromatography*. CRC Press, Cleveland.

Electronic components and circuits

Ahmed, H., and Spreadbury, P. J. (1984) *Analogue and digital electronics for engineers*. Cambridge University Press, Cambridge.

Baliga, B. Jayant and Chen, Dan Y. (1985) *Power transistors: device design and application*. IEEE, Piscataway, N.J.

Baliga, B. J. (1987) *Modern power devices*. John Wiley, Chichester, U.K.

Barna, A. and Porat, D. I. (1987) *Integrated circuits in digital electronics*. John Wiley, Chichester, U.K.

Barna, A. and Porat, D. I. (1988) *Operational amplifiers*. 2nd ed. John Wiley, Chichester, U.K.

Bird, B. M. (1983) *An introduction to power electronics*. John Wiley, Chichester, U.K.

Bishop, George Daniel. (1981) *Electrical and electronic circuits and systems*. Macmillan, London.

Bonebrear, Robert L. (1987) *Practical techniques of electronic circuit design*. 2nd ed. John Wiley, Chichester, U.K.

Boylestead, Robert L. and Nashelsky, L. (1987) *Electronic devices and circuit theory*. 4th rev. ed. Prentice-Hall, Englewood Cliffs, N.J.

Clayton, George Burbridge. (1979) *Operational amplifiers*. 2nd ed. Butterworths, London.

Clayton, George Burbridge. (1983) *Operational amplifier experimental manual*. Butterworths, London.

Davies, Rex Mountford. (1979) *Power diode and thyristor circuits*. Peregrinus for the IEE, Stevenage, U.K.

Deboo, Gordon J. and Burrows, C. N. (1977) *Integrated circuits and semiconductor devices: theory and application.* 2nd ed. McGraw-Hill, New York.

Dewan, S. B. and Straughen, A. (1975) *Power semiconductor circuits.* Wiley, New York.

Elmasry, M. I. (1981) *Digital MOS integrated circuits.* IEEE, Piscataway, N.J.

Faulkenberry, Luces M. (1982) *An introduction to operational amplifiers with linear integrated circuit applications.* 2nd rev. ed. Wiley, New York.

Floyd, Thomas L. (1983) *Essentials of electronic devices.* Merrill, Columbus, Ohio.

Floyd, Thomas L. (1984) *Electronic devices.* Merrill, Columbus, Ohio.

Floyd, Thomas L. (1987) *Electric circuit fundamentals.* Merrill, Columbus, Ohio.

Floyd, Thomas L. (1987) *Electronics fundamentals: circuits, devices and applications.* Merrill, Columbus, Ohio.

Floyd, Thomas, L. (1988) *Principles of electric circuits.* 3rd ed. Merrill, Columbus, Ohio.

Frederiksen, Thomas M. (1988) *Intuitive operational amplifiers: from basics to useful applications.* Rev. ed. McGraw-Hill, New York.

Frederiksen, Thomas M. (1989) *Intuitive analog electronics: from electron to Op amp.* McGraw-Hill, New York.

General Electric (U.S.A.) Semiconductor Products Department (1967). *Silicon controlled rectifier manual.* 4th ed. General Electric, New York.

Graeme, Jerald G. (1987) *Applications of operational amplifiers: third generation techniques.* McGraw-Hill, New York.

Graeme, J. G. (1977) *Designing with operational amplifiers: applications, alternatives.* McGraw-Hill, New York.

Gray, Paul R. *et al.* (1980) *Analog MOS integrated circuits.* IEEE, Piscataway, N.J.

Grebene, Alan B. (1978) *Analog integrated circuits.* IEEE, Piscataway, N.J.

Gregorian, Roubik, and Temes, Gabor C. (1980) *Analog MOS integrated circuits for signal processing.* Wiley-Interscience, New York.

Gyugyi, L. and Pelly, B. R. (1976) *Static power frequency changers: theory, performance, and applications.* Wiley, New York.

Honeycutt, Richard A. (1988) *Op amp and linear integrated circuits.* Delmar.

Horowitz, P. and Hill, W. (1980) *The art of electronics.* Cambridge University Press, Cambridge.

Hufault, John R. (1986) *Op amp network design.* John Wiley, Somerset, N.J.

Jacob, J. Michael. (1988) *Industrial control electronics: applications and design.* Prentice-Hall, Englewood Cliffs, N.J.

Jones, Martin Hartley. (1985) *A practical introduction to electronic circuits.* 2nd ed. Cambridge University Press, Cambridge.

Kalvoda, Robert. (1975) *Operational amplifiers in chemical instrumentation.* E. Horwood; Chichester, U.K., Halsted Press, New York.

Kosow, Irving L. (1988) *Circuit analysis.* Wiley, New York.

Kosow, Irving L. (1988) *Electronic circuits and solid state devices.* Wiley, New York.

Markus, John. (1988) *Essential circuits reference guide.* McGraw-Hill, New York.

Master handbook of 1001 practical electronic circuits: solid-state edition. Rev. ed. 1988. Tab. Bks., Blue Ridge Summit, Pa.

Meyer, Robert G. (1978) *Integrated-circuit operational amplifiers.* IEEE, Piscataway, N.J.

Moore, Brian and Donaghy, John. (1986) *Operational amplifier circuits.* Pitman, Melbourne.

Morley, M. S. (1988) *The digital IC handbook.* Tab. Bks., Blue Ridge Summit, Pa.

O'Dell, T. H. *Electronic circuit design.* (1988) Cambridge University Press, Cambridge.

Pelloso, P. (1987) *Practical digital electronics.* John Wiley, Chichester, U.K.

Pelly, B. R. (1971) *Thyristor phase-controlled converters and cycloconverters: operation, control and performance.* Wiley-Interscience, New York.

Pelly, B. R. (1987) *Power FET handbook.* Wiley, New York.

Ramamoorty, M. (1978) *An introduction to thyristors and their applications.* Macmillan, London.

Ramshaw, Raymond Southern. (1975) *Power electronics: thyristor controlled power for electric motors.* Chapman and Hall, London.

Roberge, James K. (1975) *Operational amplifiers: theory and practice.* Wiley, New York.

Roth, Charles H. (1985) *Fundamentals of logic design.* 3rd ed. West Pub. Co., St Paul, New York.

Rutkowski, George B. (1980) *Solid state electronics.* 2nd ed. Bobbs-Merrill, New York.

Smeaton, W. (ed.) (1987) *Switchgear and control handbook,* 2nd ed. McGraw-Hill, New York.

Smith, Ralph J. (1984) *Circuits, devices and systems: a first course in electrical engineering.* 4th ed. Wiley, New York.

Smith, Ralph J. (1988) *Electronics: circuits and devices.* 3rd ed. McGraw-Hill, New York.

Sparkes, John J. (1969) *Transistor switching and sequential circuits.* Pergamon Press, Oxford.

Taylor, P. D. (1987) *Thyristor design and realization.* John Wiley, Chichester, Eng.

Wait, J. V., Huelsman, L. P. and Korn, G. A. (1975) *Introduction to operational amplifier theory and applications.* McGraw-Hill, New York.

Weaver, Graham George. (1970) *Electric controls.* Morgan-Grampian, West Wickham, Eng.

Weiske, Wolfgang. (1978) *How the thyristor works.* Siemens Aktiengesellschaft, Heyden, Berlin.

Electric motors and their control

Bose, Bimal K. (1981) *Adjustable speed AC drive systems.* IEEE, Piscataway, N.J.

Bose, Bimal K. (1988) *Microcomputer control of power electronics and drives.* IEEE, Piscataway, N.J.

Chalmers, B. J. (1987) *Electric motor handbook.* Butterworths, London.

Electro-craft Corporation. (1977) *DC motors, speed controls, servo systems: an engineering handbook.* 3rd ed. Pergamon Press, Oxford.

Kenjo, Takashi and Nagamori, S. (1985) *Permanent-magnet and brushless DC motors.* Clarendon Press, Oxford.

Kosow, Irving L. (1973) *Control of electric machines.* Prentice-Hall, Englewood Cliffs, N.J.

Leonhard, Werner. (1985) *Control of electrical drives.* Springer-Verlag, Berlin, New York.

Lightband, D. A. (1970) *Direct current traction motor: its design and characteristics.* Business Books, London.

Lythall, Reginald Tarlton. (1971) *AC motor control: a guide to the basic methods of starting, controlling, sequencing and protecting AC induction motors.* Iliffe, London.

McIntyre, R. L. (1974) *Electric motor control fundamentals.* 3rd ed. McGraw-Hill, New York.

Moberg, Gerald A. (1987) *AC and DC motor control.* Wiley, New York.

Murphy, John M. D. (1973) *Thyristor control of A.C. motors.* Pergamon Press, Oxford.

Ramshaw, Raymond Southern (1973) *Power electronics: thyristor controlled power for electric motors.* Chapman and Hall, London.

Say, Maurice George. (1976) *Alternating current machines.* Wiley, New York.

Shepherd, W. and Hulley, L. N. (1987) *Power electronics and motor control.* Cambridge University Press, Cambridge.

Fluid power components and systems

Burrows, Clifford Robert. (1972) *Fluid power servomechanisms.* Van Nostrand Reinhold, New York.

Dransfield, Peter (1981) *Hydraulic control systems: design and analysis of their dynamics.* Springer-Verlag, New York.

Edwards, Harry J. (1980) *Automatic controls for heating and air-conditioning: pneumatic-electric control systems.* McGraw-Hill, New York.

Esposito, Anthony (1980) *Fluid power with applications.* Prentice-Hall, Englewood Cliffs, N.J.

Fawcett, John Reginald. (1970) *Hydraulic servo-mechanisms and their applications.* Trade and Technical Press, Morden, U.K.

Goodwin, Alfred Bernard. (1976) *Fluid power systems: theory, worked examples and problems.* Macmillan, London.

Hedges, Charles S. (1982–1984) *Industrial fluid power. (3 vols.)* 3rd ed. Wormack Educational Pubs., Dallas.

Lambeck, Raymond P. (1983) *Hydraulic pumps and motors: selection and application for hydraulic power control systems.* M. Dekker, New York.

McCloy, D. and Martin, H. R. (1980) *Control of fluid power.* 2nd ed. E. Horwood, Chichester, Eng.

Paterson, E. B. (1979) *Practical pneumatics: an introduction to low cost automation.* McGraw-Hill, Auckland, N.Z.

Pippenger, John J. (1984) *Hydraulic valves and controls: selection and application.* M. Dekker, New York.

Pippenger, John L. and Hicks, Tyler G. (1979) *Industrial hydraulics.* 3rd ed. McGraw-Hill, New York.

Principles and theory of pneumatics. (1980) Trade & Technical Press, Morden, U.K.

Prokes, Josef. (1977) *Hydraulic mechanisms in automation.* Elsevier, Amsterdam.

Rohner, Peter. (1984) *Industrial hydraulic control: a textbook for fluid power technicians.* AE Press, Melbourne.

Stewart, Harry L. and Philbin, Tom (1976). *Pneumatics and hydraulics*. 4th ed. Macmillan, New York.

Stewart, Harry L. and Storer, John M. (1973) *ABC's of hydraulic circuits*. H.W. Sams, Indianapolis.

Stewart, Harry L. and Storer, John M. (1980) *Fluidpower*. 3rd ed. Bobbs-Merrill, New York.

Stewart, Harry L. (1984) *Pneumatics and hydraulics*. T. Audel, Indianapolis.

Viersma, T. J. (1980) *Analysis, synthesis and design of hydraulic servosystems and pipelines*. Elsevier Scientific, Amsterdam.

Waller, William Frederick (1970) *Fluid control*. Morgan Grampian, West Wickham, Eng.

Wolansky, William D. and Akers, Arthur. (1988) *Modern hydraulics: the basics*. Merrill, Columbus, Ohio.

Wolansky, William D., Nagohosian, John and Henke, Russell W. (1986) *Fundamentals of fluid power*. Waveland Press, Prospect Heights, Ill.

Final control elements (non-electrical)

Beard, Chester S. (1969) *Final control elements: valves and actuators*. Rimbac, Philadelphia.

British Valve Manufacturers Association. (1980) *Valve users manual: a technical reference book on industrial valves for the control of fluids*, edited by J. Kempley. Mechanical Engineering Pubs., London.

Hutchinson, J. W. and Merrick, A. R. (1976) *ISA handbook of control valves*. 2nd ed. Instrument Society of America, Pittsburgh.

Instrument Society of America. (1985) *Final control elements*. The Society, Pittsburgh.

Pearson, George Harold (1979) *Valve design*. Mechanical Engineering Pubs., London.

Stewart, Harry L. and Philbin, Tom. (1984) *Pumps*. 4th ed. Macmillan, New York.

Wood, P. (1988) *Development in valves and actuators for fluid control: 2nd International Conference, Manchester, Eng, 1988*. Springer-Verlag, London.

Mechanical system elements

Chironis, Nicholas P. (1965) *Mechanisms, linkages and mechanical controls*. McGraw-Hill, New York.

Dijksman, E. A. (1976) *Motion geometry of mechanisms*. Cambridge University Press, Cambridge.

Drago, Raymond J. (1988) *Fundamentals of gear design*. Butterworths, London.

Mabie, Hamilton Horth and Owirk, Fred W. (1987) *Mechanisms and dynamics of machinery*. 4th ed. Wiley, New York.

Index